Ulrico Peckelsen

Objective Tyre Development

Definition and Analysis of Tyre Characteristics
and Quantification of their Conflicts

Karlsruher Schriftenreihe Fahrzeugsystemtechnik
Band 57

Herausgeber
FAST Institut für Fahrzeugsystemtechnik
Prof. Dr. rer. nat. Frank Gauterin
Prof. Dr.-Ing. Marcus Geimer
Prof. Dr.-Ing. Peter Gratzfeld
Prof. Dr.-Ing. Frank Henning

Das Institut für Fahrzeugsystemtechnik besteht aus den Teilinstituten Bahnsystemtechnik, Fahrzeugtechnik, Leichtbautechnologie und Mobile Arbeitsmaschinen.

Eine Übersicht aller bisher in dieser Schriftenreihe erschienenen Bände finden Sie am Ende des Buchs.

Objective Tyre Development

Definition and Analysis of Tyre Characteristics and Quantification of their Conflicts

by
Ulrico Peckelsen

KIT Scientific Publishing

Dissertation, Karlsruher Institut für Technologie
KIT-Fakultät für Maschinenbau

Tag der mündlichen Prüfung: 15. Mai 2017
Referenten: Prof. Dr. rer. nat. F. Gauterin, Prof. Dr. M. Gobbi

Impressum

KIT Scientific Publishing

Karlsruher Institut für Technologie (KIT)
KIT Scientific Publishing
Straße am Forum 2
D-76131 Karlsruhe

KIT Scientific Publishing is a registered trademark
of Karlsruhe Institute of Technology.
Reprint using the book cover is not allowed.

www.ksp.kit.edu

Print on Demand 2017 – Gedruckt auf FSC-zertifiziertem Papier

ISSN 1869-6058
ISBN 978-3-7315-0713-0
DOI 10.5445/KSP/1000073428

Vorwort des Herausgebers

Die Fahrzeugtechnik ist gegenwärtig großen Veränderungen unterworfen. Klimawandel, die Verknappung einiger für Fahrzeugbau und –betrieb benötigter Rohstoffe, globaler Wettbewerb, gesellschaftlicher Wandel und das rapide Wachstum großer Städte erfordern neue Mobilitätslösungen, die vielfach eine Neudefinition des Fahrzeugs erforderlich machen. Die Forderungen nach Steigerung der Energieeffizienz, Emissionsreduktion, erhöhter Fahr- und Arbeitssicherheit, Benutzerfreundlichkeit und angemessenen Kosten finden ihre Antworten nicht aus der singulären Verbesserung einzelner technischer Elemente, sondern benötigen Systemverständnis und eine domänenübergreifende Optimierung der Lösungen.

Hierzu will die Karlsruher Schriftenreihe für Fahrzeugsystemtechnik einen Beitrag leisten. Für die Fahrzeuggattungen Pkw, Nfz, Mobile Arbeitsmaschinen und Bahnfahrzeuge werden Forschungsarbeiten vorgestellt, die Fahrzeugsystemtechnik auf vier Ebenen beleuchten: das Fahrzeug als komplexes mechatronisches System, die Fahrer-Fahrzeug-Interaktion, das Fahrzeug in Verkehr und Infrastruktur sowie das Fahrzeug in Gesellschaft und Umwelt.

Die Fahrzeugentwicklung nutzt immer stärker virtuelle Prototypen um schneller, flexibler und kostengünstiger auf die Anforderungen des Markts reagieren zu können. Bei dem Bauteil Reifen, der einen erheblichen Einfluss auf viele Gebrauchseigenschaften des Fahrzeugs hat, ist dies bislang noch nicht so gut gelungen, wie bei anderen Fahrzeugteilsystemen. Dies hat seine Ursache vor allem in den komplexen Vorgängen im Rollkontakt, in dessen Einfluss auf die Dynamik der Reifenstruktur und des Luftraums im Reifeninneren sowie in den von vielen Einflussgrößen abhängen Materialeigenschaften. Physikalische Reifenmodelle sind dadurch sehr umfangreich und rechenintensiv. Ein FEM-Modell eines Reifens mit seinem kompletten mehrlagigen Aufbau kommt selbst in vereinfachter geometrischer Darstellung schnell auf 10^6 Freiheitsgrade. Dabei werden fast immer vereinfachte Materialgesetzte verwendet.

Um die Modelle dennoch im Entwicklungsprozess nutzen zu können, sind sie vielfach auf die Beschreibung einzelner Reifen-Gebrauchseigenschaften zuggeschnitten. Ein einfaches Modell für die Beschreibung aller Gebrauchseigenschaften des Reifens existiert nicht. Dieses wäre aber hilfreiche, um die zahlreichen Zielkonflikte quantifizieren zu können, die in der Reifenentwicklung aufgrund des starken Einflusses des Reifens auf viele Gebrauchseigenschaften des Fahrzeugs bestehen.

Hier setzt die Arbeit von Herrn Peckelsen an, in der er die Kette von Konstruktionsparametern des Reifens über Kenngrößen des Gesamtreifens bis hin zu im Fahrmanöver am Fahrzeug erfassbaren objektiven und subjektiven Größen des Fahrverhaltens einbezieht. Über ein vereinfachtes physikalisches Modell ermittelt er die Zusammenhänge zwischen Reifengestaltungsparametern und Reifenkenngrößen, den Einfluss der Reifenkenngrößen auf das Fahrzeugverhalten bildet er über statistisch ermittelte Zusammenhänge ab. Aus der Betrachtung der gesamten Kette heraus quantifiziert er die Zielkonflikte zwischen vier ausgewählten Gebrauchseigenschaften des Fahrzeugs hinsichtlich des Einflusses der Reifengestaltung.

Frank Gauterin

Karlsruhe, 24.7.2017

Objective Tyre Development Definition and Analysis of Tyre Characteristics and Quantification of their Conflicts

Zur Erlangung des akademischen Grades
Doktor der Ingenieurwissenschaften
Der Fakultät für Maschinenbau
Karlsruher Institut für Technologie (KIT)

genehmigte
Dissertation

von

M. Sc. Ulrico Peckelsen

Tag der mündlichen Prüfung:	15.05.2017
Hauptreferent:	Prof. Dr. rer. nat. F. Gauterin
Korreferent:	Prof. Dr. M. Gobbi

Abstract

The present thesis focuses on the tyre, especially on its influence on four requirements of the vehicle development, namely power loss, lateral dynamics, ride comfort and interior noise. The objective of the thesis is the quantification of conflicts between four selected requirements considering the physical constraints given by the tyre.

The method proposed in the present thesis is based on a set of functional tyre characteristics (FTCs), a physical tyre model and a procedure for identifying and quantifying the conflicts. The FTCs are objective quantities that can be derived from tyre simulation or tyre measurement (e.g. vertical stiffness); they are a "common language" for communicating tyre characteristics. The physical tyre model and the proposed procedure allow to evaluate conflicts as a function of geometrical and material properties of the tyre.

The method contributes in reducing time and costs of the tyre development; moreover, the know-how generated through the performed objectivation studies and sensitivity analyses supports decision-making during the virtual design of tyres as well as vehicle architecture, axle kinematics and wheel-suspension system characteristics.

Kurzfassung

Der Fokus der vorliegenden Dissertation liegt auf dem Reifen, insbesondere auf seinem Einfluss auf vier Anforderungen in der Fahrzeugentwicklung, nämlich der Rollwiderstandsverlustleistung, der Querdynamik, der Vertikaldynamik und dem Innengeräusch des Fahrzeugs. Ziel der Dissertation ist die Quantifizierung der Zielkonflikte zwischen den Anforderungen. Diese Zielkonflikte entstehen aufgrund der durch den Reifen gegebenen physikalischen Randbedingungen.

Die vorgestellte Methode fußt auf einem Set an funktionalen Reifeneigenschaften (functional tyre characteristics, FTCs), einem physikalischen Reifenmodell und einem Prozess zur Identifizierung und Quantifizierung der Zielkonflikte. Die FTCs sind objektive Eigenschaften, deren Werte von einem virtuellen Reifenmodell oder einer Reifenmessung abgeleitet werden können (z. B. eine Vertikalsteifigkeit); sie stellen folglich eine „gemeinsame Sprache" für die Kommunikation der Reifeneigenschaften dar. Das physikalische Reifenmodell und der vorgestellte Prozess ermöglichen es die Zielkonflikte und deren Abhängigkeit von geometrischen und Materialeigenschaften des Reifens zu berechnen und visualisieren.

Die Methode trägt dazu bei, Zeit und Kosten der Reifenentwicklung zu reduzieren. Darüber hinaus unterstützen die durch Objektivierungsstudien und Sensitivitätsanalysen gewonnen Erkenntnisse die Konzeptauswahl und –gestaltung von Reifen, sowie die virtuelle Grundauslegung der Fahrzeug-, Achs, und Federdämpfungsarchitektur in der frühen Entwicklungsphase.

Acknowledgement

The present thesis addresses several topics and different requirements of the tyre: from statistic methods for DoE to physical modelling of tyres, from power loss and lateral dynamics to ride comfort and interior noise phenomena. Therefore, strong collaborations with different departments and experts were needed. In the following, I would like to thank all those colleagues and friends who played an important role during the thesis period, well-knowing that a lot more contributed in shaping it.

First of all, I would like to thank the colleagues of the tyre development department, especially those who participated in long-lasting studies at the test track and at the driving simulator: C. Vogt, T. Beck, G. Mülbl, H. Dreier, P. Wagner and R. Rapp as well as D. Bode and R. Högl. The opportunity to share knowledge and skills concerning vehicle dynamics and tyre development was of great help!

Furthermore, a big thank goes to the colleagues of the simulation and the NVH departments. Dr. F. Niedermeier and Dr. M. Zimmermann had been perfect feedback partners concerning virtual tyre development; N. Reissig, T. Heinker, Dr. M. Brandstätter, Dr. M. Klein and Dr. L. Witta supported the planning and the evaluation of studies concerning interior noise and vehicle comfort.

A doctorate thesis about tyres could not be performed without a strong collaboration with tyre manufactures: R&D projects, analyses of series tyres, measurements at special tyre test benches and their in depth knowledge of tyre production and development enriched the thesis, especially in those topics dealing with conflicts evaluation between tyre requirements.

A special thank goes to M. Pehlke and B. van der Meer, the two department leader, who promoted and supported the thesis over its all period, increasing confidence and acceptance for the topic at BMW.

Last but not least, a special thanks goes to my doctoral thesis supervisor Prof. Dr. rer. nat. F. Gauterin, who contributed in shaping the topic and the target of the thesis enriching it. The excellent collaboration, especially in the compilation of the research publications, and the constant eye kept on the overall goal of the thesis helped in creating an innovative and coherent research study.

I would also like to thank Dr. H.-J. Unrau for the comprehensive feedback, especially regarding rolling resistance, and the great suggestions given in the final composition of the thesis.

Finally, a great thank goes to my two tutors with whom I spent most time meditating on the thesis: T. D'Avanzo and Dr. M. Bullinger. Their uncomplicated and straightforward tutorship permitted a great cooperation throughout the period of the thesis: they gave brilliant suggestions and ideas finding always time for feedback!

Karlsruhe, December 2016

Ulrico Peckelsen

Contents

1 Introduction

Today, original equipment manufacturers (OEMs) have to face a highly competitive market. On the one hand, they have to meet customer needs and satisfy a constant demand for innovation; on the other hand, they have to achieve legal targets and reduce product development time and costs. Moreover, passenger vehicles are characterized by several attributes, among others, fuel consumption, comfort and driving performance: they shape the character of the vehicle and affect the perceived quality of the product representing an important differentiating element for OEMs.

The present thesis focuses on the tyre, especially on its influence on four requirements of the vehicle development, namely power loss, lateral dynamics, ride comfort and interior noise. The objective of the thesis is the quantification of conflicts between four selected requirements considering the physical constraints given by the tyre. The physical behaviour of the tyre is here described by functional tyre characteristics (FTC). Focusing on the FTCs has one essential advantage: the FTCs are objective quantities that can be derived from a virtual tyre model or a tyre measurement; they are a "common language" for communicating tyre characteristics. The method proposed in the present thesis contributes in reducing time and costs of the tyre development; moreover, the know-how generated through the performed objectivation studies and sensitivity analyses supports decision-making during the virtual design of tyres as well as vehicle architecture, axle kinematics and wheel-suspension system characteristics.

For the requirements concerning power loss, lateral dynamics, ride comfort and interior noise, first, a set of objective manoeuvre criteria (OMC) is defined; then, the most significant FTCs are identified. The OMC allow an objective description of the target behaviour of the full-vehicle, on test track or in simulation; the FTCs allow a unique description of the tyre characteristics. They can be derived from simulation models, as well as from test bench measurements. Correlation and sensitivity analysis are implemented for identifying the type and the strength of the main relations between OMC and FTCs. The output of these analyses are subsets containing the FTCs that mainly influence the targets defined by the OMC of each requirement. Finally, a physical tyre models is introduced: it describes the tyre dynamics and, subsequently, the FTCs, as a function of the tyre design parameters and allows to quantify the conflicts between the requirements concerning power loss, lateral dynamics, ride comfort and interior noise.

The present thesis is structured in ten sections. In the introductory section the motivation, the objectives of the thesis and its structure are presented. In the second section the state of the art concerning the requirements power loss, lateral dynamics, ride comfort and interior noise are presented. In the third section the choice of models and methods is motivated and their development are introduced. Furthermore, a definition of "tyre design parameter" (TDP), "functional tyre characteristic" (FTC), "objective manoeuvre criterion" (OMC) and "subjective manoeuvre index" (SMI) is given. In sections four to seven the requirements power loss, lateral dynamics, ride comfort and inter noise are analysed separately: measurements, objectivation studies and global sensitivity analyses necessary to identify OMC and related FTCs are presented. In section eight the developed physical tyre model is introduced. This model is used to identify and quantify the conflicts between the requirements: the results of the correlation analyses are presented in section nine. In the last section the main results are summarised and the added scientific value is pointed out. The thesis is concluded with suggestions for future developments.

2 State of the Art

In this section, first, the choice of the requirements power loss, lateral dynamics, ride comfort and interior noise is motivated. Then, the state of the art concerning their objectivation and the influence tyres have on them is presented.

The present thesis focuses on the quantification of conflicts between four selected requirements, namely power loss, lateral dynamics, ride comfort and interior noise. In order to quantify the conflicts it is necessary to identify the relevant tyre characteristics and analyse their influence on the requirements. Although the selected requirements are not the only ones tyres have to satisfy (e.g. tyre uniformity, drill torque for parking manoeuvres, tyre abrasive wear, ...), they are ranked among the most significant ones from both a customer and a legislative perspective. Objectivation studies prove a strong correlation to customer driving experience, while legislature introduces more and more standards to regulate fuel consumption, driving safety and noise pollution.

Analysis concerning the power loss of tyres started in the beginning of the 20th century (Wormeley & Holt, 1922). In the 21st century, environmental awareness and stricter CO_2 regulations (European Union, 2009) raised attention on rolling resistance and spurred the introduction of innovative solutions to reduce power loss. Tyres, being the ultimate component in the well-to-wheel chain, are a crucial leverage (ERDA Rolling Resistance Advisory Committee, 1977). Analyses show, that there is a linear relation between rolling resistance and fuel consumption (Bradley & Delaval, 2013): rolling resistance is estimated to contribute with 16 % to the overall power loss of vehicles (Schulze, Bolz, Strübel, & Wies, 2010), share that may vary due to the observed driving cycle (e.g. mainly highway use) and vehicle concept (e.g. electric cars). Due to its contribution to CO2 levels a tyre label measuring amongst other rolling resistance is introduced (European Union, 2009).

The handling characteristics of vehicles are strongly influenced by tyres. Their geometrical and material properties play a major role in defining steering wheel feedback, driving feeling and vehicle stability (Zomotor, Braess, & Rönitz, 1997) (Zomotor, Braess, & Rönitz, 1998). Lateral dynamics is evaluated directly by customers, but also by test drivers in press-tests and, concerning safety, by the severe lane change manoeuvre (ISO 3888-2, 2011) based on the so-called Swedish "moose test".

Road asperities, e.g. transversal ribs, gullies and bumps, generate vibrations, in longitudinal, lateral and vertical direction (Guiggiani, 2014). In given frequency ranges, humans are particularly sensitive to these vibrations (BS 6841, 1987) (ISO 2631, 1997). As the tyre is one of the main filters between road excitation and passenger it represents a crucial leverage for improving ride comfort and, subsequently, the driving pleasure of the passengers.

Similarly to ride comfort, tyres characterize the interior noise of vehicles. Tyres play a major role in filtering road excitations due to road asperities and road roughness: "quiet" tyres reduce interior noise and assure a comfortable and relaxing ride (Genuit, 2010). Tyres are also responsible for exterior noise (Sandberg & Ejsmont, 2002): coast by test (ISO 362, 2009) should quantify the vehicle's overall exterior noise and the contributions given by tyres (European Union, 2009).

2.1 Rolling Resistance

In this section, first, an overview of common formulations describing rolling resistance is given. Then, the state of the art concerning the influence tyres have on it is presented.

2.1.1 Formulations for Rolling Resistance

A first modelling of the rolling resistance is proposed in 1785 by Coulomb. He assumes that the resistance to the rolling of tyres is caused by rolling friction due to road asperities the tyre must surmount. A possible mathematical formulation for the rolling resistance force F_r is (Coulomb, 1785):

$$F_r = e_C \frac{F_z}{h_C} \cong e_C \frac{F_z}{R} \tag{2.1}$$

where e_C and h_C are the horizontal and vertical Coulomb's distances between wheel hub and road obstacle, F_z is the tyre load and R the tyre radius.

Later, Reynolds showed that road asperities are not the only cause of rolling resistance: the deflection of the tyre plays an important role, too (Reynolds, 1874). Today, several formulations exist to describe the rolling resistance of tyres (Schuring, 1977): it can be represented as a force, a torque, an energy loss (Schuring, 1976) or a power loss (Holt & Wormeley, 1922).

A common formulation for the total rolling resistance force F_r is the sum of tyre bending resistance $F_{r,bend}$, tyre air resistance $F_{r,air}$, tyre frictional resistance $F_{r,fr}$ (micro-slips in longitudinal and lateral direction), road resistance $F_{r,road}$, resistance due to slip angle $F_{r,slip}$, and resistance due to bearing friction and residual braking $F_{r,fb}$ (Heißing & Ersoy, 2011):

$$F_r = F_{r,bend} + F_{r,air} + F_{r,fr} + F_{r,road} + F_{r,slip} + F_{r,fb} \tag{2.2}$$

2.1.2 Influence of Tyre Geometry, Tyre Material and Tyre Operating Conditions on Rolling Resistance

The rolling resistance is influenced by several parameters (LaClair, 2006): amongst others, tyre geometry (Clark, 1977), tyre operating conditions (Clark, 1978) (Peckelsen & Gauterin, 2013), and road surface texture (DeRaad, 1977).

One of the first attempts of defining analytically the influence of the tyre geometry on rolling resistance is presented in (Evans, 1960) and then developed by (Clark, 1977). The analytical results are based on tyres for commercial vehicles. It can be stated, that rolling efficiency (the reference tyre has a rolling efficiency of 1.00) increases moderately with tyre diameter (see Figure 2.1) and decreases slightly with tyre width. Moreover, for a given inflation pressure, tyre load and tyre vertical deflection η (expressed as ratio of the sidewall height) there is a combination of tyre diameter and tyre width that minimises rolling efficiency.

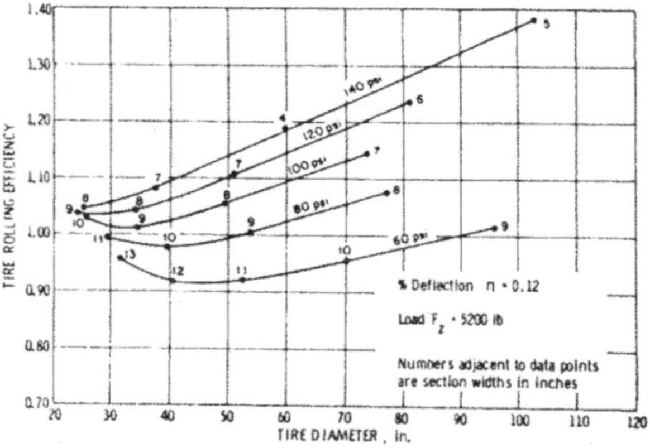

Figure 2.1: Influence of tyre diameter at constant load and aspect ratio (Clark, 1977)

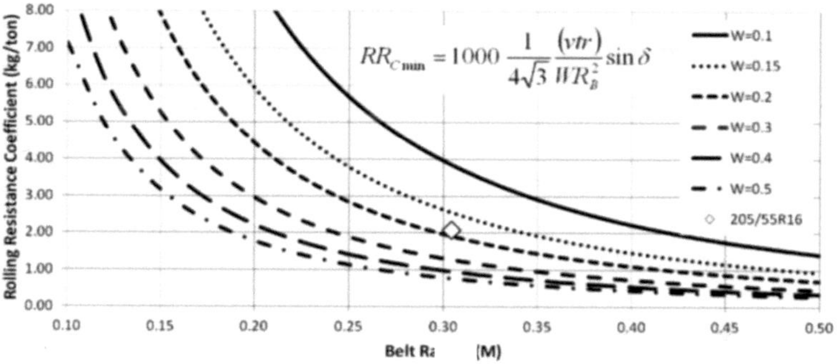

Figure 2.2: Influence of tyre diameter and width at constant load and aspect ratio (Rhyne & Cron, 2012)

More recent analytical studies show similar results: in Figure 2.2 the influence of tyre diameter (here referred to as belt radius R_B) and tyre width W is shown (here depicted as positive for rolling resistance).

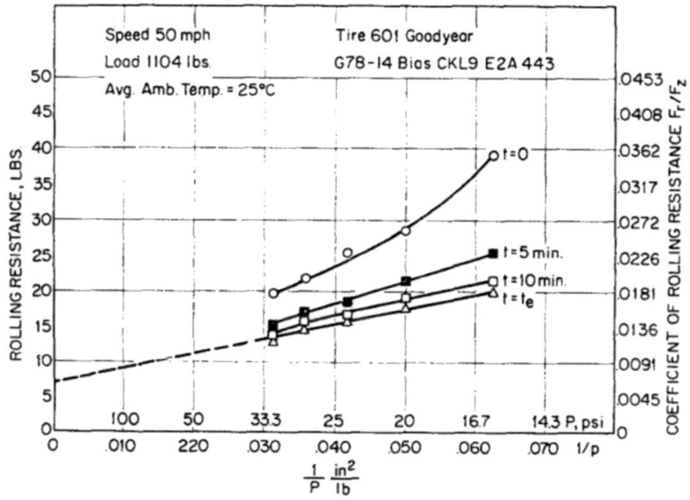

Figure 2.3: Rolling resistance versus reciprocal of inflation pressure after different measurement times (i.e. temperatures) (Clark, 1978)

Concerning the tyre operating conditions, measurements are presented in (Clark, 1978). In Figure 2.3 shows the relations between rolling resistance, inflation pressure and temperature. Rolling resistance decreases with temperature: longer warm-up phases reduce rolling resistance. The relation between rolling resistance and inflation pressure is hyperbolic for “warm” tyres (~40° at 80 km/h free rolling straight-forward driving), and strongly hyperbolic for “cold” tyres (ambient temperature).

The influence of tyre operating conditions concerning slip angle, camber angle, tyre load and trajectory velocity is presented in (Peckelsen, D'Avanzo, Bode, Brenker, & Gauterin, 2013) and (Peckelsen & Gauterin, 2013). The analyses are based on a design of experiments (DoE) composed of 50 different measurement conditions on an outer drum test bench (ISO 28580, 2009). The ranges of the operating conditions are derived from data collected by vehicle customers (see Table 2.4). The results show, that total rolling resistance force is almost independent from trajectory velocity (see Figure 2.5) and camber angle. On the contrary, it is a linear function of tyre load and a parabolic function of slip angle (see Figure 2.6).

Table 2.4: Operating conditions

Operating condition	Symbol	Range
Trajectory velocity	v_x	30 to 150 km/h
Slip angle	α	$\pm 2.5°$
Camber angle	γ	$\pm 3.5°$
Tyre load	F_z	1000 to 5000 N

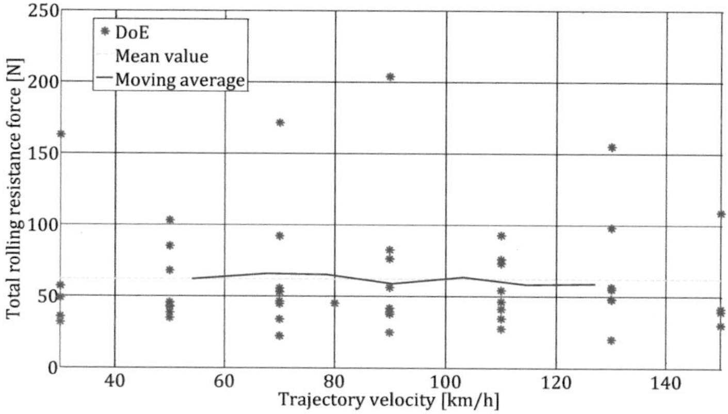

Figure 2.5: Influence of trajectory velocity on total rolling resistance force

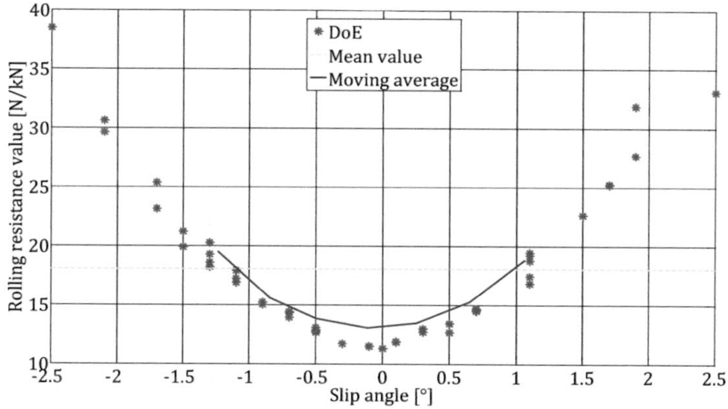

Figure 2.6: Influence of slip angle on rolling resistance value (total rolling resistance force divided by tyre load)

2.2 Lateral Dynamics

In this section, first, an overview of the relevant manoeuvres for lateral dynamics and their objective manoeuvre criteria is given. Then, the state of the art concerning the influence tyres have on it is presented.

2.2.1 Manoeuvres to Evaluate Lateral Dynamics

The characteristics concerning the lateral dynamics of vehicles can be subdivided into four main categories (Zomotor, Braess, & Rönitz, 1997): driving in a curve, driving straight-forward, transient response to single steering inputs and transient response to multiple inputs. These characteristics can be evaluated according to several manoeuvres: examples are given in Table 2.7.

Table 2.7: A possible classification of the characteristics of lateral dynamics

Characteristics	Examples of manoeuvres
Driving in a curve	Quasi steady-state cornering; Power off while cornering; Braking while cornering
Driving straight-forward	Aquaplaning; μ-split breaking; Side wind sensibility; Steering response
Transient response to single steering inputs	Step steer; Lane change
Transient response to multiple inputs	Continuous sine steer; Step steer while accelerating;

Table 2.8: OMC for lateral dynamics

Manoeuvre	Input	References
RAST	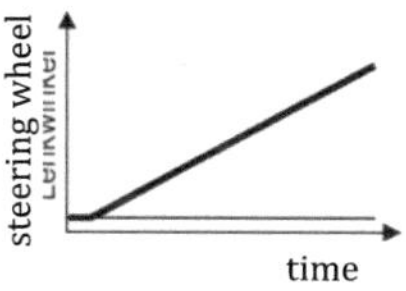	(Weir & DiMarco, 1978) (Rompe & Ehlich, 1978) (Strange, 1982) (Dibbern, 1992) (Fuchs, 1993) (Redlich, 1994) (Riedel & Arbinger, 1997) (Chen & Crolla, 1998) (Henze, 2004) (Harrer, 2007) (Botev, 2008) (Niedermeier, 2015)
CSST CSW CSSW	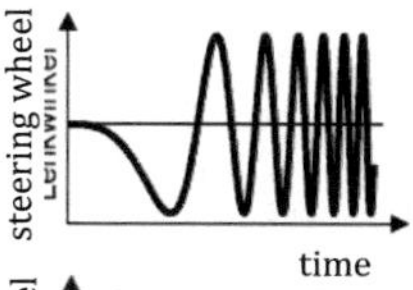 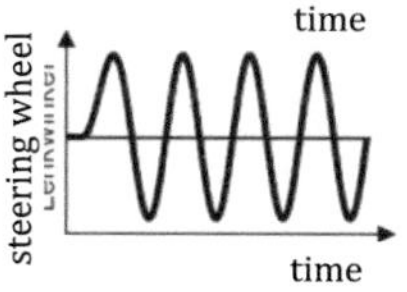	(Deppermann, 1989) (Dibbern, 1992) (Farrer, 1993) (Redlich, 1994) (Gies & Marusic, 2000) (Henze, 2004) (Dettki, 2005) (Harrer, 2007) (Schimmel, 2010) (Niedermeier, 2015)
STST	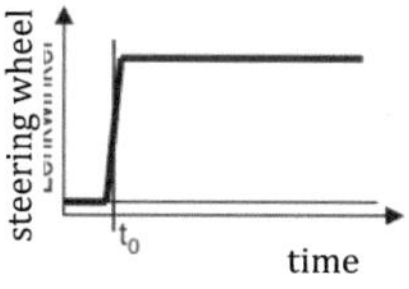	(Weir & DiMarco, 1978) (Kudritzki, 1989) (Dibbern, 1992) (Fuchs, 1993) (Redlich, 1994) (Riedel & Arbinger, 1997) (Zschocke, 2009) (Decker, 2009)

Each manoeuvre can be evaluated subjectively and objectively. The first attempts to explain subjective evaluations with objective manoeuvre criteria[1] (OMC) started in 1973 by (Bergman, 1973). An overview about subjective evaluations is given by (Heißing & Brandl, 2002) (Gutjahr, 2014); an exhaustive review concerning objective criteria is presented in (Botev, 2008) (Decker, 2009) (Huneke, 2012) (Schimmel, 2010) (Gutjahr, 2014).

2.2.1.1 Ramp Steer

The “quasi steady-state cornering” (QSSC) is probably the oldest and most common manoeuvre used to identify the steady-state behaviour of vehicles: it is performed on a circular path of a given radius ($30\,m$ to $110\,m$) increasing quasi-statically the driving velocity. In simulation, the “ramp steer” (RAST) is preferred, as, contrary to the QSSC, it is an open-loop manoeuvre: subsequently, it allows a faster simulation and more accurate evaluation of the vehicle behaviour. The RAST is performed at constant velocity ($30\,km/h$ to $100\,km/h$); the input is a quasi-statically increasing steering wheel angle. Both manoeuvres allow the evaluation of the characteristics of sideslip angle β, lateral acceleration a_y and steering wheel torque M_H as a function of the steering wheel angle δ_H. Typical OMC (see references in Table 2.8) are gain factors:

$$\frac{X}{\delta_H} \qquad X \in [\beta, a_y, M_H] \qquad \textbf{2.3}$$

and gradients:

$$\frac{\partial X}{\partial \delta_H} \qquad X \in [\beta, a_y, M_H] \qquad \textbf{2.4}$$

[1] See section 3.1.3 for the definition of OMC.

2.2.1.2 Continuous Sine Steer

The “continuous sine steer” (CSST) is performed at constant velocity ($80\ km/h$ to $120\ km/h$); the input is a sinusoidal steering wheel angle with increasing frequency and constant steering wheel angle amplitude. Similar manoeuvre are the “continuous sine wave” (CSW) and the “continuous sine sweep” (CSSW): both are performed at constant velocity ($80\ km/h$ to $120\ km/h$) with a sinusoidal steering wheel angle that is characterized either by constant frequency and amplitude (CSW), or by constant frequency and increasing amplitude (CSSW). All three manoeuvres allow the evaluation of the characteristics of yaw velocity $\dot{\psi}$, lateral acceleration a_y and steering wheel torque M_H as a function of the steering wheel angle input δ_H. The CSST focuses on the frequency response functions; typical OMC (see references in Table 2.8) are the eigenfrequency-gain:

$$\max\left(\frac{X}{\delta_H}\right) \qquad X \in [\dot{\psi}, a_y] \qquad \mathbf{2.5}$$

and the phase shift φ at $-45°$ or the equivalent time delay Δt:

$$\begin{matrix} \varphi_{X,45°} \\ \Delta t_{eq,X,45°} \end{matrix} \qquad X \in [\dot{\psi}, a_y] \qquad \mathbf{2.6}$$

The CSW allows the calculation of the hysteretic behaviour of the vehicle response ($\dot{\psi}, a_y$) and the steering wheel torque (M_H) in relation to the steering input. Typical OMC (see references in Table 2.8) focus on the description of the “ellipse” describing the hysteretic cycle: e.g. the mean slope of the major axis of the ellipse and the width of the ellipse (comparable with the minor axis).

Finally, the CSSW focuses on single frequencies (e.g. $0.5\ Hz$, $1Hz$, $1.5\ Hz$) and evaluates the variation of the vehicles response due to the steering amplitude (and subsequently the lateral acceleration). Typical OMC (see references in Table 2.8) are gain factors:

$$\frac{X}{\delta_H} \qquad X \in [\dot{\psi}, a_y, M_H] \qquad \textbf{2.7}$$

and gradients:

$$\frac{\partial X}{\partial \delta_H} \qquad X \in [\dot{\psi}, a_y, M_H] \qquad \textbf{2.8}$$

2.2.1.3 Step Steer

The "step steer" (STST) is performed at constant velocity. The input is a step steering wheel angle, characterized by a "rising phase" (a fraction of a second) and by a "holding phase", which allows the vehicle to reach steady-state conditions. The STST allows the evaluation of the characteristics of sideslip angle β, yaw velocity $\dot{\psi}$ and lateral acceleration a_y. Typical objective criteria (see references in Table **2.8**) are gain factors at steady-state condition "ss":

$$\left.\frac{X}{Y}\right|_{ss} \qquad X, Y \in [\beta, \dot{\psi}, a_y] \qquad \textbf{2.9}$$

time delays between vehicle response and steering input:

$$\Delta t_{X-\delta_H} \qquad \Delta t_{X-Y} \qquad X, Y \in [\beta, \dot{\psi}, a_y] \qquad \textbf{2.10}$$

and the peak of the overshoot (oscillations) before reaching steady-state condition ("ss"):

$$\frac{X_{max} - X_{ss}}{X_{ss}} \qquad X \in [\beta, \dot{\psi}, a_y] \qquad \textbf{2.11}$$

2.2.2 Influence of Tyre Geometry, Tyre Material and Tyre Operating Conditions on Lateral Dynamics

In order to evaluate the influence tyre geometrical and material properties as well as tyre operating conditions have on lateral dynamics a set of characteristic values (CVs) is introduced (Niedermeier, Peckelsen, & Gauterin, 2013). These CVs describe the tyre longitudinal and lateral dynamics of the tyre and allow to evaluate effects of the tyre design and of the tyre operating conditions on the vehicle lateral dynamics.

2.2.2.1 Characteristic Values for Lateral Dynamics

In (Niedermeier, Peckelsen, & Gauterin, 2013) a set of 58 CVs for longitudinal and lateral dynamics is presented: 14 CVs for the longitudinal force, 22 for the lateral force and 22 for the self-aligning torque allow a unique and exhaustive description of the tyre longitudinal and lateral dynamics. Their influence is evaluated for different manoeuvres concerning longitudinal and lateral dynamics.

The CVs of each force and torque can be subdivided into static and dynamic CVs. The static CVs concern the description of force and torques over longitudinal slip and slip angle. The dynamic CVs describe the relaxation lengths. Concerning the static CVs, four main CVs describe the longitudinal force, five main CVs the lateral force and five main CVs the self-aligning torque (see Table 2.9). Additional CVs describe the linear variation of the main CVs due to camber angle, tyre load and longitudinal slip. For example, the variation of the cornering stiffness K_y due to tyre load is:

$$K'_{y,0,F_z} = \frac{\Delta K_y}{\Delta F_z} \cdot 100 = \frac{\left(K_y|_{F_{z,max}} - K_y|_{F_{z,min}}\right)}{F_{z,max} - F_{z,min}} \cdot 100 \qquad \textbf{2.12}$$

where K_y and F_z are respectively the cornering stiffness and the tyre.

Table 2.9: The five main CVs for lateral force

Name	Formulation	Symbol	Unit
Lateral force at zero slip angle	$F_y\|_0$	$F_{y,0}$	$[N]$
Cornering stiffness	$\left.\frac{\partial F_y}{\partial \alpha}\right\|_{0^\circ}$	$K_{y,0}$	$[N/°]$
Maximal lateral force	$\max(F_y)$	$F_{y,max}$	$[N]$
Position of maximal lateral friction	$\alpha\|_{F_{y,max}}$	$\alpha_{F_{y,max}}$	[°]
Lateral force at 15° slip angle	$F_y\|_{15^\circ}$	$F_{y,lim}$	$[N]$

The variation ranges for tyre load, camber angle and longitudinal slip are fixed ranges (see Table 2.10): they are chosen to generate a realistic variation for normal driving conditions. Concerning the dynamic CVs, two CVs describe respectively the longitudinal and lateral relaxation length and their variaiton due to tyre load (see Table 2.11).

Table 2.10: Ranges for the calculation of the variaitons of the CVs for lateral force

Variation	Symbol	Range
Longitudinal slip	κ	$[-0.1, 0.1]$
Camber angle	γ	$[0°, -4°]$
Tyre load	F_z	$[2000N, 6000N]$

Table 2.11: The two CVs for lateral relaxation length

Name	Formulation	Symbol	Unit
Lateral relaxation length	$\sigma_y\vert_0$	$\sigma_{y,0}$	$[m]$
Variation of lateral relaxation length due to tyre load	$\frac{\Delta\sigma_y}{\Delta F_z}\cdot 100$	$\sigma'_{y,0,F_z}$	$[m/N]$

2.2.2.2 Influence of Characteristic Values on Lateral Dynamics

In (Peckelsen, 2012) and (Niedermeier, 2015) global sensitivity analyses between the CVs and OMC are performed in order to identify the relevant CVs. The influence of the CVs on the OMC is quantified by Sobol's total sensitivity index (see section 3.5.2.1) presented exemplarily in Table 2.12 for the ramp steer (OMC 4 and OMC 5 are omitted due to confidentiality).

Further analyses, concerning also the continuous sine steer and the step steer, are presented in (Niedermeier, 2015): twelve CVs describing the lateral force, its variation due to camber angle and tyre load, as well as the self-aligning torque are identified as most significant for the lateral dynamics.

Table 2.12: Global sensitivity analysis between CVs and OMC of a ramp steer

	$a_{y,max}$	$\frac{\beta}{\delta_H}\Big\vert_{4\,m/s^2}$	$\frac{\partial\beta}{\partial\delta_H}\Big\vert_{4\,m/s^2}$	OMC 4	OMC 5
$K_{y,0}$	0	**0.75**	0.33	0.32	0
$F_{y,max}$	**0.64**	0	0.01	0.01	0.06
$\alpha_{F_{y,max}}$	0.01	0	0.06	0.03	0.03
$F_{y,lim}$	0.18	0.04	0.04	0.01	0

2.3 Ride Comfort

In this section, first, the human perception of vibration is described. Then, the state of the art concerning the manoeuvres used to evaluate ride comfort and the influence tyres have on it is presented.

2.3.1 Vehicle Vibration Sources

The main excitation sources of the ride comfort of a vehicle are road texture, tyre non-uniformities and power unit (Mitschke & Wallentowitz, 2003) They induce mechanical vibrations of different vehicle components (see Figure 2.13). The passenger feels these vibrations through the passenger seat and, concerning the driver only, through the steering wheel.

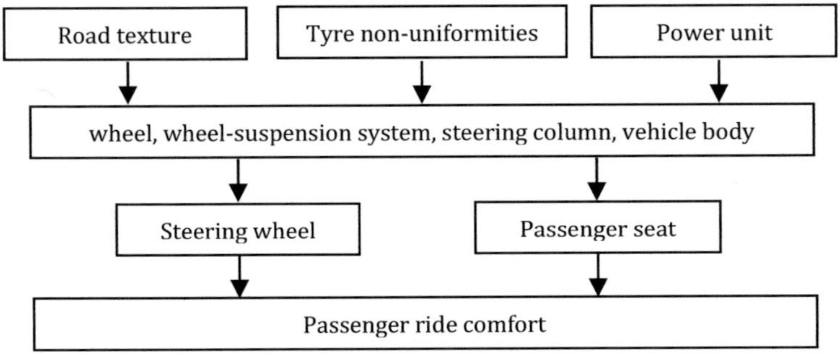

Figure 2.13: Ride comfort sources

Of course, also driver steering inputs, as well as braking inputs, can excite ride comfort phenomena (e.g. a continuous sine steering input excites the roll eigenmode of a vehicle).

For ride comfort due to road excitations, both tyre (Hilscher, 2008) (Michelin, 2005) and wheel-suspension system (Troulis, 2002) represent crucial components: depending on the frequency, they contribute in transmitting or isolating road excitations influencing ride comfort. According to seismic theory, below its eigenfrequency a harmonic oscillator generates a response to an input signal (e.g. road excitation) according to its mass, damping and stiffness characteristics; on the contrary, if excited above approximately 1.5 times its eigenfrequency it uncouples seismically output from input. Moreover, through its mass, stiffness and damping characteristics it influences frequency and amplitude of its eigenmode. Typical eigenfrequencies of the vehicle body are between 1 and 2 Hz, of the passenger seat between 2 and 3 Hz and of the unsprung masses (wheel-suspension system) between 10 and 20 Hz (Mitschke & Wallentowitz, 2003). So, tyres have two main contributions: transmitting road excitations, i.e. forces and accelerations at the wheel-hub, and varying the frequency and amplitude of the wheel-suspension system's eigenmode. In (Fülbier, 2001) a method is presented, to analyse the complete transmission path from the road to the passenger.

2.3.2 Human Perception of Vibration

The human perception of whole-body vibration is studied exhaustively in (Dupuis & Zerlett, 1984), (Bobbert, 1988), (Simic, 1970) (Dupuis & Hartung, 1972), (Christ, 1973) and (Griffin & Fothergill, 1977). According to these studies, the human perception of vibration depends on frequency, amplitude, direction and position of the stress. The two main standards for its evaluation are (BS 6841, 1987) and (ISO 2631, 1997).

For humans, vibrations are particularly uncomfortable around 5 Hz. In fact, the majority of the eigenfrequencies of the human body lies below 30 Hz: motion sickness is caused by low frequency vibrations perceived by the inner ear and the eyes, while impacts and oscillations are felt by larger parts of the human body (e.g. stomach, legs, arms). Some examples are given in Table 2.14 according to research studies presented in (Dupuis & Zerlett, 1984), (Parson & Kraemer, 1987) and (Heißing & Brandl, 2002).

Concerning the amplitude of the vibrations, guidelines for the thresholds for the human perception of acceleration are enlisted in (VDI 2057 Blatt 1, 2002): accelerations greater than 1 m/s^2 are evaluated as uncomfortable (Michelin, 2005).

Table 2.14: Eigenfrequencies in vertical direction of parts of the human body when sitting in a vehicle's passenger seat

Part of the human body	Eigenfrequency
legs	1 to 4 Hz
backbone	3 to 6 Hz
chest / body	3 to 6 Hz
stomach	4 to 7 Hz
shoulder	5 to 10 Hz
arm	12 to 35 Hz
head	15 to 30 Hz

According to position and direction of the stress, two different weights are applied: a weighting curve covering the whole frequency spectrum and a weighting factor for the direction. These weights change also according to the type of excitation and are defined in (BS 6841, 1987) and (ISO 2631, 1997). So, local indices measuring the accelerations at feet, seat and backrest can be summed to an overall discomfort index: the method presented in Figure 2.15 is based on (BS 6841, 1987).

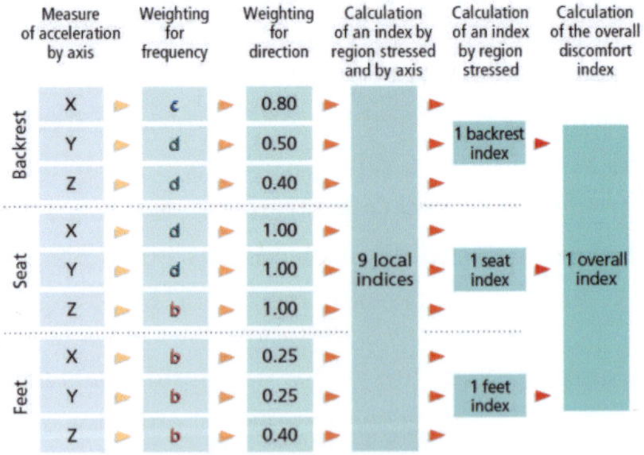

Figure 2.15: Stages in calculating a discomfort index (Michelin, 2005) (BS 6841, 1987)

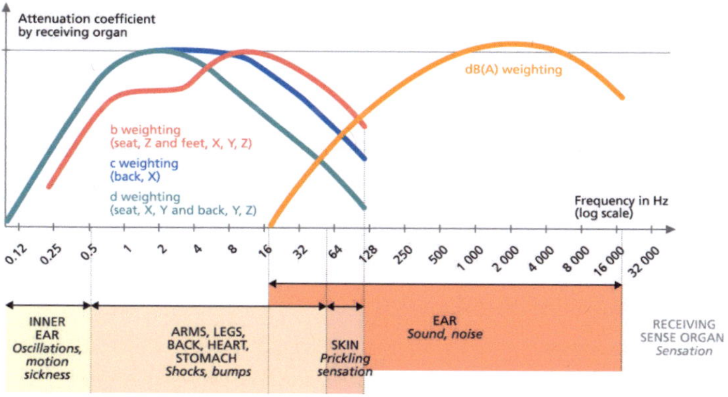

Figure 2.16: Perception of vibration as a function of frequency, position and direction (Michelin, 2005)

In Figure 2.16 an overview of frequency ranges of the organs is represented. The diagram can be subdivided in three main areas: frequencies below 20 Hz concern pure vibration phenomena, while frequencies above 100 Hz concern pure acoustic phenomena (see diagram (Hilscher, 2008)). The frequency range in between, from 20 to 100 Hz, characterizes the transition from vibration to sound (Hieronimus, 1990).

2.3.3 Manoeuvres to Evaluate Ride Comfort

The evaluation of ride comfort is based on spectra representing forces and accelerations measured, for example, at the wheel hub and at the vehicle centre of mass. These spectra can be derived from vehicle measurements or simulations. When possible, frequency response functions (FRFs) are preferred, because they take into account the spectrum of the excitation (e.g. road amplitudes or induced test bench forces).

Manoeuvres for the evaluation of ride comfort are performed on irregular roads driving straight-forward at constant velocity (30 km/h to 80 km/h): the irregularities can be harmonic excitations (VDI 2057 Blatt 1, 2002), stochastic excitations (Hennecke, 1995) and single obstacles (Cucuz, 1993). To identify the spectrum of the vehicle (Hilscher, 2008) suggests to use harmonic excitations: they allow to scan accurately the vehicle reaction at defined frequencies and amplitudes in the frequency range of ride comfort, typically 0 to 30 Hz.

Ride comfort phenomena concern the eigenmodes of rolling φ, pitching ϑ and bouncing z_s of the sprung mass (vehicle body), as well as the eigenmodes of longitudinal x_u, lateral y_u and vertical acceleration z_u of the unsprung masses (wheel-suspension system). The eigenfrequencies of the vehicle body lie between 0.5 and 2 Hz; those of the wheel-suspension system between 10 and 15 Hz (Michelin, 2005). The evaluation of ride comfort is often represented by an integral value of a given frequency range of the FRF (Rericha, 1986) (Bobbert, 1988) (Mitschke, Cucuz, & Hennecke, 1995) (Griffin, 2012). Typical integral values for the vehicle FRF are:

$$\sqrt{\int_{f_1}^{f_2} \left(\frac{X}{z_e}\right)^2 df} \qquad X \in [\ddot{\varphi}, \ddot{\vartheta}, a_{z,s}] \qquad \mathbf{2.13}$$

where z_e is the road excitation, $a_{z,s}$ the vertical acceleration of the sprung masses and the frequencies f_1 and f_2 are chosen around 0.5 Hz and 2 Hz in order to characterize the first eigenmode of the vehicle bouncing, pitching or rolling. Integral values for the FRF of the wheel-suspension system are:

$$\sqrt{\int_{f_3}^{f_4} \left(\frac{X}{z_e}\right)^2 df} \qquad X \in [a_{x,u}, a_{y,u}, a_{z,u}] \qquad \mathbf{2.14}$$

where $a_{\dots,u}$ are the acceleration of the unsprung masses and the frequencies f_3 and f_4 are chosen around 10 Hz and 20 Hz in order to described the contribution given by the eigenmode of wheel and wheel-suspension system.

2.3.4 Influence of Tyre Geometry, Tyre Material and Tyre Operating Conditions on Ride Comfort

Ride comfort is influenced by several parameters: amongst others, tyre material and geometrical properties (Hilscher, 2008) (Fülbier, 2001), tyre operating conditions (Fülbier, 2001), and road surface texture (Fülbier, 2001).

One of the most important parameter for ride comfort is the tyre vertical stiffness, in particular the dynamic stiffness of a rolling tyre (Hilscher, 2008). The vertical stiffness of tyres can be measured under static or dynamic conditions, for a standing or a rolling tyre. The static vertical stiffness is evaluated at constant tyre load and deflection, while the dynamic vertical stiffness is measured under a periodic excitation with constant amplitude and frequency at an average tyre load. Both, static and dynamic vertical stiffness, can be measured for a standing and rolling tyre: at low speed stiffness (and also damping) decreases with velocity (Jianmin, Gall, & Zuomin, 2001). In Figure 2.17 measurements of the characteristics of the dynamic vertical stiffness at different velocities is shown.

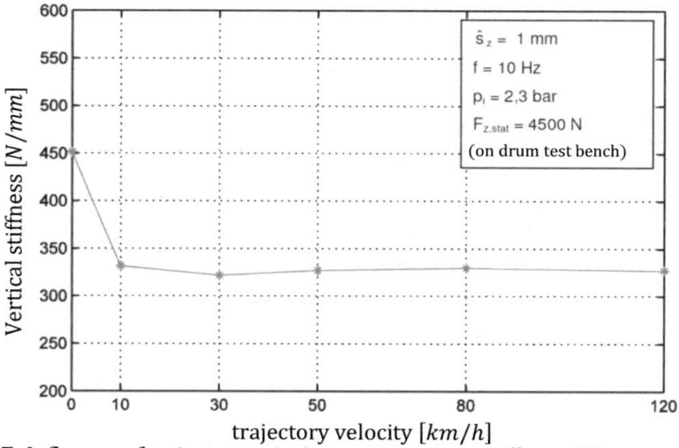

Figure 2.17: Influence of trajectory velocity on the vertical stiffness (Niemeyer, Eckstein, Kessen, Klein, & Wegener, 2011)

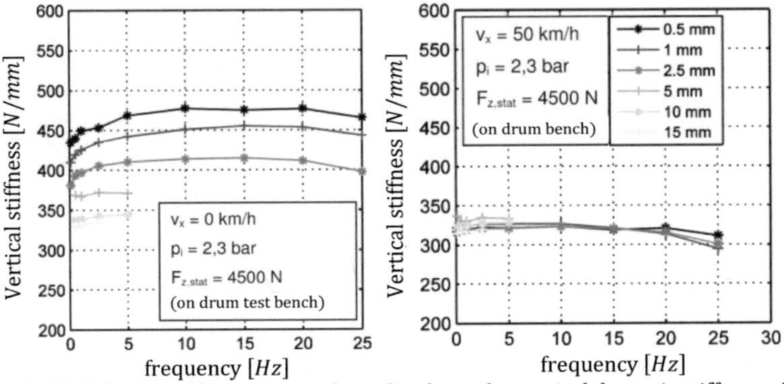

Figure 2.18: Influence of frequency and amplitude on the vertical dynamic stiffness of a standing and a rolling tyre (Niemeyer, Eckstein, Kessen, Klein, & Wegener, 2011)

Measurements prove, that the dynamic stiffness of a rolling tyre is almost independent from the excitation amplitude (see Figure 2.18), but slightly dependent on the tyre load (see Figure 2.19).

Concerning material and geometrical properties, all components influencing vertical stiffness have an influence on ride comfort. Examples are the sidewall stiffness and the belt bending stiffness, which determines the tyre enveloping stiffness (Hilscher, 2008).

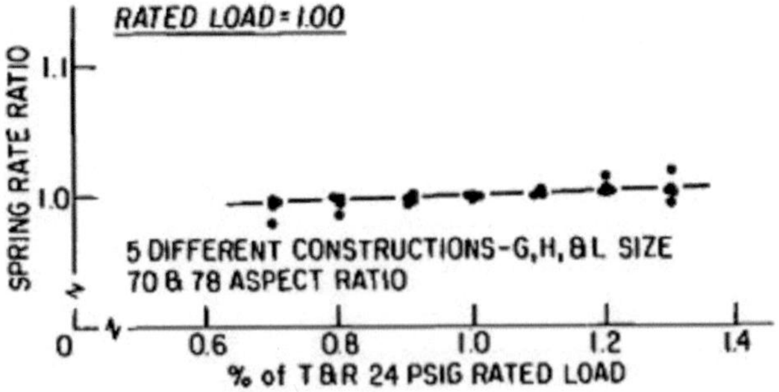

Figure 2.19: Influence of tyre load on the spring rate ratio (normalised vertical dynamic stiffness) (Pottinger & Yager, 1986)

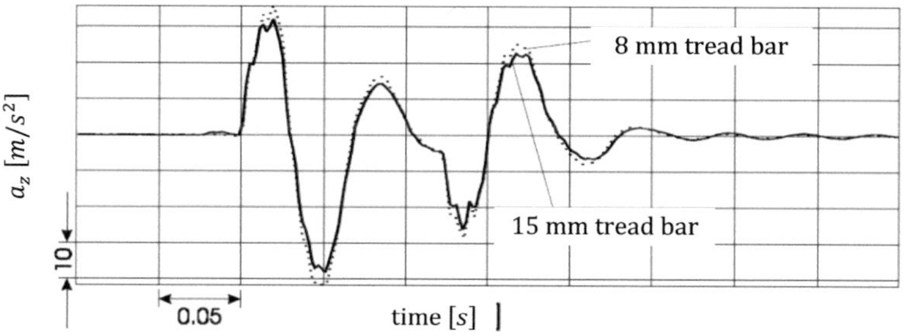

Figure 2.20: Vertical acceleration at the wheel hub as a function of the tread bar height (Fülbier, 2001)

Concerning tread height, simulations show, that it affects slightly ride comfort (see Figure 2.20). Instead, the damping characteristics of tyres are negligible compared to those of the wheel-suspension system (Mitschke & Wallentowitz, 2003).

Concerning the operating conditions, (Fülbier, 2001) identifies the tyre inflation pressure as most important parameter for improving ride comfort. Simulations show, that low inflation pressure decreases forces and accelerations at the wheel hub, improving the experienced ride comfort (see Figure 2.21).

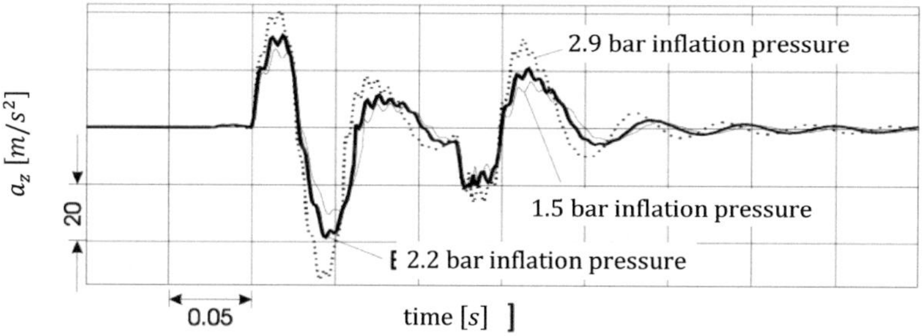

Figure 2.21: Vertical acceleration at the wheel hub as a function of the tyre inflation pressure (Fülbier, 2001)

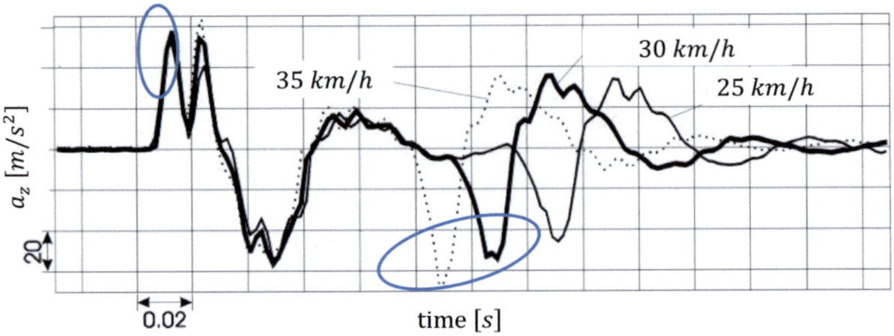

Figure 2.22: Vertical acceleration at the wheel hub as a function of the trajectory velocity (Fülbier, 2001)

Two further operating conditions are tyre load and trajectory velocity. Tyre load varies very slightly the vertical stiffness of the tyre: the higher the tyre load, the higher the vertical stiffness (see Figure 2.19). Trajectory velocity changes (indirectly) the damping properties of the wheel-suspension system: the higher the velocity the harder the damping (see Figure 2.22). Subsequently, an increase of tyre load or trajectory velocity influences negatively ride comfort, although their effects are negligible if compared to inflation pressure or obstacle type and geometry.

Concerning road irregularities, (Fülbier, 2001) analyses the influence of the obstacle type and geometry. Both show an influence on the subjective evaluation (see Figure 2.23): higher and sharper obstacles (e.g. kerbstones) decrease ride comfort.

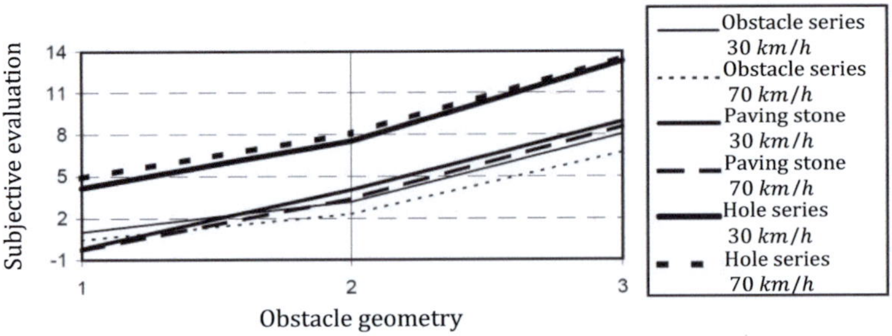

Figure 2.23: Vertical acceleration at the wheel hub as a function of three different obstacle geometries (Fülbier, 2001)

2.4 Acoustics

In this section, first, a classification of vehicle noise sources is given. Then, the human perception of sound is described. Finally, the state of the art concerning the manoeuvres used to evaluate interior noise and the influence tyres have on it are presented.

2.4.1 Vehicle Noise Sources

Overall vehicle noise can be subdivided into wind-turbulence noise, power unit noise and tyre-road noise as shown in Figure 2.24 (Sandberg & Ejsmont, 2002). Wind turbulence noise is generated at high velocities by the air turbulence around the vehicle. Power unit noise is produced by all mechanical units of the vehicle that take part in propulsion, except tyres. Tyre-road noise concerns noise generated by the interaction of tyre and road surface. Overall vehicle noise can also be subdivided into interior noise and exterior noise, i.e. into noise heard respectively inside and outside the vehicle (Sandberg & Ejsmont, 2002). As shown in Figure 2.25 interior noise is generated by the contributions of structure-borne and airborne noise of wind turbulence, power unit and tyre-road interaction: it concerns all acoustic phenomena between 30 Hz and 20 kHz.

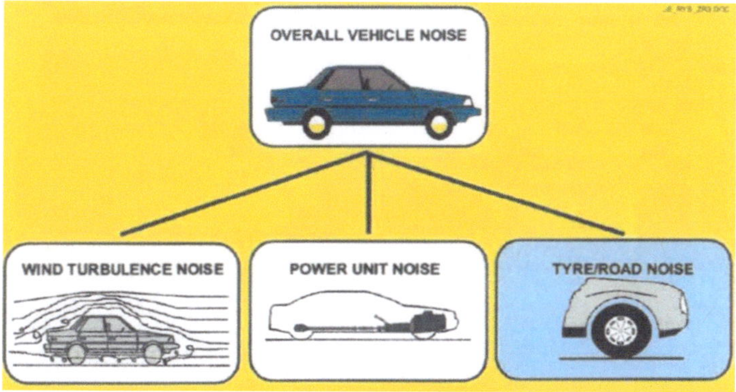

Figure 2.24: Noise emission sources for a passenger car (Sandberg & Ejsmont, 2002)

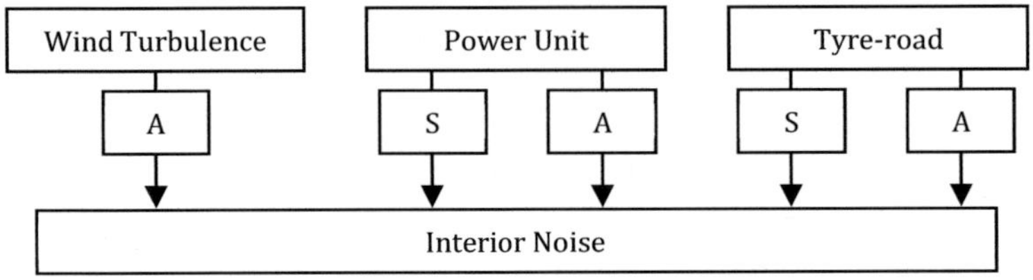

Figure 2.25: Contributions to interior noise (S: structure-borne, A: air-borne)

A review of experimental studies concerning the contributions of single parts of the vehicle, e.g. exterior mirror, windshield wiper, antenna, A-pillar, vehicle under-body, wheelhouse, to the generation of wind turbulence noise are presented in (Helfer, 2010).

The three main contributions to interior noise, given by wind turbulence, power unit and tyre/road, are dependent from engine load and vehicle velocity. As shown in Figure 2.26 at low engine loads the contribution of wind turbulence and tyre-road interaction are dependent from the vehicle velocity (Zeller, 2012). With increasing velocity wind turbulence gains importance. Experimental studies show that at ca. 70 km/h the noise generated by wind turbulence is comparable to that generated by tyre-road interaction (Dobrzynski, 1983); above 130 km/h wind turbulence is the primary noise source of a vehicle (Riegel & Wiedemann, 2003).

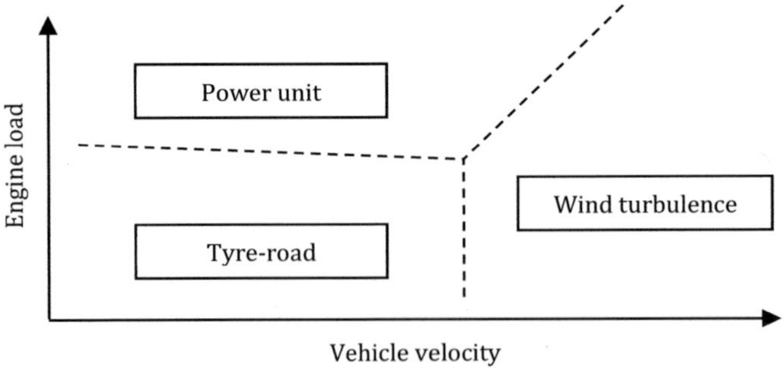

Figure 2.26: Main contributions to interior noise in relation to engine load and vehicle velocity (Brandstätter, 2013)

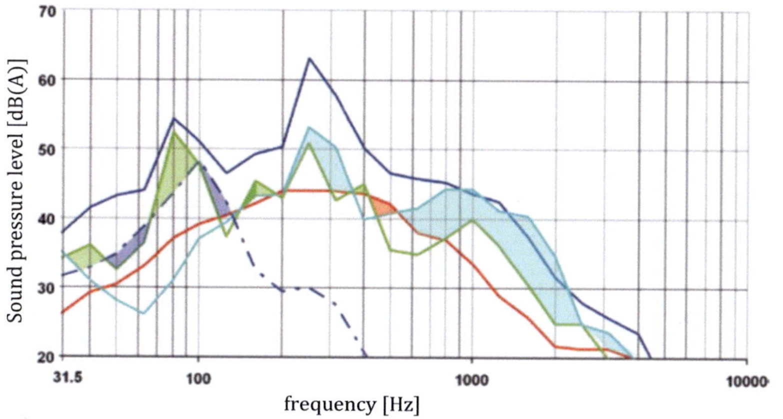

Figure 2.27: Measurements of the main contributions to the interior noise of a middle-class vehicle at 70 km/h on asphalt track (Riegel & Wiedemann, 2008)

Measurements concerning the tyre-road structure-borne and air-borne noise are presented in (Riegel & Wiedemann, 2008). In Figure 2.27 measurements for a middle-class limousine are exemplarily shown: below 100 Hz interior noise is dominated by the power unit and the tyre-road structure-borne noise; above 100 Hz the tyre air-borne noise as well as wind turbulence gain importance.

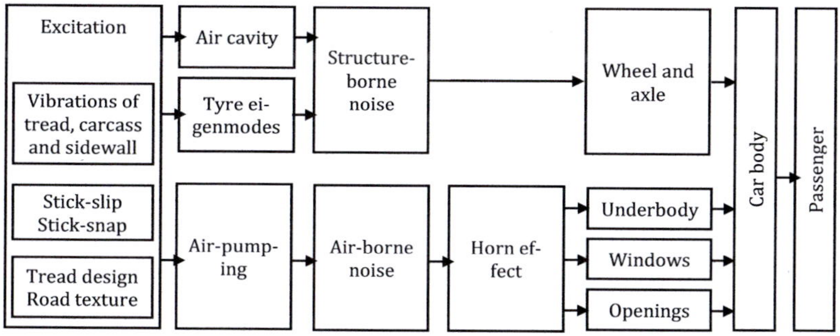

Figure 2.28: Development path of interior noise (Gauterin, 2010)

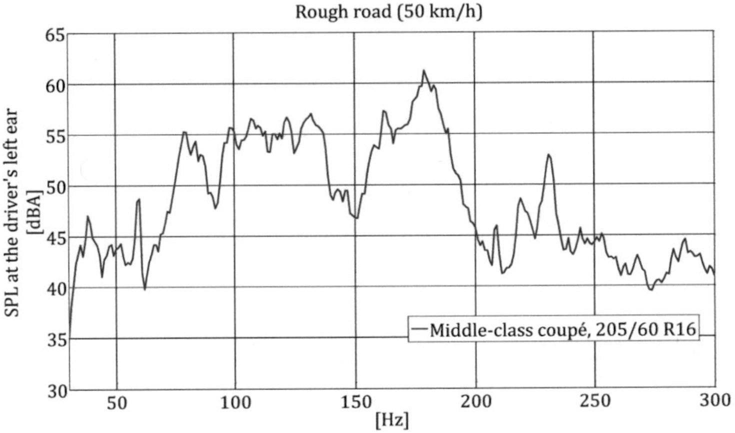

Figure 2.29: Measured sound pressure level at the driver's left ear

The present thesis focuses on interior noise caused by the tyre-road interaction. An overview of the several excitation mechanisms is given in (Sandberg & Ejsmont, 2002). The structure-borne phenomena concern vibrations of the tread, carcass and sidewall due to the tyre-road interaction and due to stick-slip and stick-snap effects, as well as the air cavity resonance, which propagates through the tyre and rim structure. The airborne phenomena include air pumping and pipe resonances: both can be amplified by horn effects and by tread design and road texture (see Figure 2.28).

Interior noise generated by tyres is typically analysed in a frequency range between 30 and 300 Hz (Gauterin & Ropers, 2005). A typical spectrum of an A-weighted sound pressure level (SPL) of interior noise is shown in Figure 2.29.

The sound pressure level (SPL) expressed in dB is (Brüel & Kjaer, 1993):

$$L_p = 10 \cdot \log\left(\frac{p}{p_{ref}}\right)^2 \qquad \textbf{2.15}$$

where p_{ref} is an standardized reference sound pressure of $20 \cdot 10^{-6}\ Pa$. The reference value is set, that $0\ dB$ correspond to the threshold of hearing. A sound is two times louder if the difference between p_1 and p_2 is approximately $3\ dB$, i.e. a SPL difference of $10\ phon$ (unit of perceived loudness level of pure tones at $10\ kHz$).

2.4.2 Human Perception of Sound

Sound is a variation of air density and propagates with a speed of about $340\ m/s$ (Sandberg & Ejsmont, 2002). The generated variations of pressure can be measured by a membrane of a microphone and are expressed in Pascal $[Pa]$. At very low frequencies sound pressure is perceived as mechanical vibration. The transition between pure sound to pure mechanical vibration occurs in the range between 20 and $100\ Hz$ (Hieronimus, 1990). Sound below $20\ Hz$ is defined as “infrasound”; above $20\ kHz$ as “ultrasound” and almost not audible by the human ear. The acoustic perception changes with age: for elderly people the highest audible frequency drops to about $5\ kHz$ (Sandberg & Ejsmont, 2002).

The common range for “audible sound” is between 20 and $20000\ Hz$: in this range the sensitivity for sound pressure is dependent from frequency. Therefore, weighting filters are developed. The two most common weighting filters are described by the A-weighting and the C-weighting curve. The formulation of the A-weighting curve in $dB(A)$ is (DIN EN 61672-1, 2003):

$$A(f) = 20 \cdot log\left[\frac{f_4^2 f^4}{(f^2 + f_1^2)\sqrt{f^2 + f_2^2}\sqrt{f^2 + f_3^2}(f^2 + f_4^2)}\right] - A_{1000} \qquad \mathbf{2.16}$$

where f is the frequency, f_1, f_2, f_3, f_4 and A_{1000} are standardised coefficients (see Table 2.30). The A-weighted SPL is:

$$L_{pA} = 10 \cdot \log\left(\frac{p}{p_{ref}}\right)^2 - A(f) \qquad \mathbf{2.17}$$

Table 2.30: Standardised coefficients for A-weighting curve and C-weighting curve

Coefficients	Value
f_1	20.60 Hz
f_2	107.7 Hz
f_3	737.9 Hz
f_4	12194 Hz
A_{1000}	0.062 dB

It is commonly used for acoustic studies: it represent well the human perception of sound (Sandberg & Ejsmont, 2002) as proven experimentally by (Fletcher & Munson, 1933) and (Robinson & Dadson, 1956). The equal-loudness-level contours of Fletcher and Munson are standardised in (ISO 226, 2003).

2.4.3 Manoeuvres to Evaluate Interior Noise

In this section, the main phenomena concerning interior noise generated by tyres are defined. The definitions are based on the descriptions presented in (Gauterin, 2010).

A possible classification of phenomena concerning vehicle interior noise is given in Figure 2.31: hum (Wummern), rumble (Donnern) and drone (Dröhnen) are examples of low frequency interior noise generated by tyres. Additional phenomena are described in (Gauterin, 2010), e.g. high rumble (rauer Ablauf), impact damping (Stoßempfindlichkeit) and sound pressure (Druckgefühl).

Hum is a low-frequency noise ($< 100\ Hz$) with modulating sound intensity and frequency caused by non-uniformity of tyres rolling on a smooth road at velocities between 30 and 80 km/h. A comprehensive study on the noise generated by hum is given in (Brandstätter, 2013). It is known that hum has a psychological influence on the driver: it generates tiredness (Genuit, 2010) and nausea (DIN 45680, 1997).

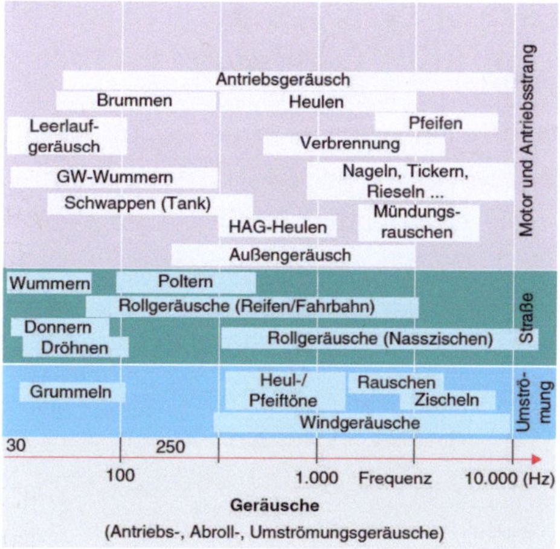

Figure 2.31: Acoustic phenomena concerning vehicle interior noise (Pletschen, 2010)

Drone is a tonal effect of eigenmodes of the vehicle generated by resonances of the complete wheel or the tread-pattern-design. It typically occurs on smooth roads at velocities between 30 and 130 km/h.

Rumble and high rumble are defined as a band-pass noises in the low (from 30 to 100 Hz) and middle frequency range (from 100 to 350 Hz). Both are caused by eigenmodes of tyres and vehicle when driving on rough asphalt at velocities between 10 and 120 km/h.

Impact damping describes the noise generated when driving up and down over cleats with a velocity of 30 to 80 km/h. An initial noise-vibration peak is followed by a slow fading in a frequency range between 3 and 130 Hz.

Sound pressure at very low frequencies and with high amplitudes is a phenomenon that is rather felt than heard. It is caused by low-frequency eigenmodes of vehicle and wheel-assembly.

2.4.4 Influence of Tyre Geometry, Tyre Material and Tyre Operating Conditions on Interior Noise

In order to objectify the influence of tyres on interior noise the frequency response function (FRF) between road excitation and wheel hub is measured. The FRF of a tyre is dependent on the type of connection of the wheel hub: on a test track the tyre is elastically connected to the vehicle through the wheel-suspension system; on a test bench the wheel hub is normally fixed. A comparison between test bench measurements and on-track measurements (also front axle to rear axle) is shown in (Sochor, 2014). Due to the big effort needed to determine the road excitation input in (Sochor, 2014) only the spectra at the wheel hub are compared and not the FRF.

The shape of the tyre's spectrum is characterized by its eigenmodes: a study based on a finite element model is presented in (Aboutorabi & Kung, 2012). The spectrum can be influence by material and geometrical properties of the tyre, as well as by operating conditions (e.g. inflation pressure, tyre load, vehicle trajectory and ambient temperature).

Concerning the material and geometrical properties, in (Aboutorabi & Kung, 2012) the influence of structural components of the tyre (amongst others: apex, bead, belt, chafer, sidewall, tread) on the frequency and generalised mass of the tyre's FRF around the 150 Hz range is quantified. The objective of the study is to identify those components that allow to reduce most the amplitude of vibrations and hence the loudness of the generated noise. The stiffness of the ply cord as well as the mass of the tread allow significant variations of interior noise. A method to predict the interior noise generated by tyres based on their FRF is presented in (Bahnert, Lienkamp, & Vogel, 2012): the interior noise is calculated multiplying the frequency response function (FRF) of the vehicle, as measured on track for a reference tyre, with the FRF of the tyre, as measured on a test bench. It is based on the assumption, that the FRF of the vehicle is independent from the tyre FRF.

Concerning the operating conditions, the effect of rotation speed on the tyre dynamic behaviour at different tyre loads is analysed experimentally and numerically in (Kindt, et al., 2013) and (Grollius, 2013). However, the assumption that at non-zero rotational speed the wave speeds of positive- and negative-going waves diverge from each other due to Coriolis accelerations is not been proven yet. Moreover, it is not shown that this effect influences interior noise. Measurements regarding the influence of trajectory velocity, inflation pressure and tyre load are presented in (Sochor, 2014). It is shown, that higher trajectory velocity and higher inflation pressure causes the forces at the wheel hub to rise and, subsequently, the sound pressure level at the driver's left ear to increase. On the contrary, variations of tyre load do not show any significant effect on interior noise, even if they change the eigenfrequency of the tyre eigenmodes.

Temperature has also an effect on interior noise. In (Anfosso-Lédée & Pichaud, 2007), linear relations between noise level (here measured as pass-by noise, which is part of the air-borne interior noise) and air temperature are observed: the higher the temperature the lower the stiffness and damping of the tyre. The influence is particularly strong for bituminous pavements ($-0.1\ dB(A)/°C$), relatively low for pavements having porosity ($-0.06\ dB(A)/°C$) and absent for cement concrete pavements.

2.5 Conflicts between the Requirements

Concerning the quantification of the conflicts, analyses seldom deal with more than two requirements at once and they are often performed at vehicle level, not at tyre level. Analyses of conflicts at tyre level are limited to statistics and design studies based on the variation of only some geometrical and material properties of the tyre (e.g. tyre contour, tread material, ...). However, no quantification at tyre level of the conflicts between the four selected requirements is given.

Concerning conflicts at vehicle level, handling and ride comfort characteristics are among the most studied ones. In (Botev, 2008) the subjective evaluation of handling and ride comfort characteristics are first analysed separately and then compared, showing that these requirements are generally in conflict. Also in (Mitschke & Wallentowitz, 2003) conflicts arising between ride comfort and driving safety are explained: wheel-suspension systems that filter better road excitations cause higher tyre-load oscillations and, subsequently, less driving safety. Possible solutions to lessen the conflict are active control systems (Botev, 2008) (Heißing & Ersoy, 2007) and optimisation of axle and wheel-suspensions system (Heißing & Ersoy, 2011). In (Heißing & Ersoy, 2007) and (Denker, 1988) conflicts between handling (or driving safety) and noise, vibration and harshness (NVH) phenomena are addressed: damping and stiffness characteristics of wheel-suspension system are enlisted as crucial parameters.

Concerning conflicts at tyre level, an overview is given in (Heißing & Ersoy, 2007). The conflict between wet grip and rolling resistance is among the most analysed ones. In the past, tyres showed strong conflict (Auto Bild, 2009), which is lessen thanks to the new materials introduced in modern tyres (Greiner & Heimann, 2013). These materials for tread and sidewalls have low loss modulus at excitation frequencies between 5 and 15 Hz (typical of the tyre rotation frequency for trajectory velocities of 10 to 30 m/s and tyre diameters of 0.6 m) and high loss modulus at frequencies between 10^3 and 10^5 Hz (typical of micro-slip phenomena that generate friction) as analysed in (Schulze, Bolz, Strübel, & Wies, 2010). In (Landwehr, 2013) new materials for tread and sidewalls are analysed in order to reduce the conflict between wet grip, rolling resistance and tyre abrasive wear. In (Venneborger, Strübel, Wies, & Wiese, 2013) the influence of the tread design as well as of tread and sidewall material properties on different requirements is analysed: the objective is the development of a tyre for electric and hybrid vehicles, that have good driving comfort and performance but higher cruising range.

3 Choice of Methods

In this section, new acronyms are defined and the choice of methods concerning vehicle simulation, design of experiments and correlation analyses is motivated. Moreover, the state of the art of the chosen methods is presented and further developments are introduced (e.g. concerning the homogeneity of design of experiments).

3.1 Definitions

In this section, the definition of four new acronyms is given: the tyre design parameter (TDP), the functional tyre characteristic (FTC), the objective manoeuvre criterion (OMC) and the subjective manoeuvre index (SMI).

3.1.1 Tyre Design Parameter

A tyre design parameter (TDP) defines a measurable or calculable value of the geometrical or the material property of the tyre. Examples are the thickness of a sidewall or the elastic modulus of the belt. A TDP can be measured on a tyre test bench or in a laboratory.

The introduction of an appropriate set of TDPs is crucial. Each TDP influences one or more FTCs and, subsequently, one or more OMC. So, the TDPs allow the quantification of conflicts between the OMC of selected requirements.

3.1.2 Functional Tyre Characteristic

A functional tyre characteristic (FTC) defines a measurable or calculable value of the tyre dynamics. Examples are the slip stiffness or the rolling resistance torque (see also (Niedermeier, Peckelsen, & Gauterin, 2013)).

A FTC can be represented by an analytical expression. Moreover, each FTC is a function of the TDPs and influences one or more OMC.

An appropriate set of FTCs allows a unique description of the tyre dynamics without using characteristic curves and without defining the tyre's geometrical and material properties. It can be used to compare two tyres or define the target dynamics of a tyre.

3.1.3 Objective Manoeuvre Criterion

An objective manoeuvre criterion (OMC) defines a measurable or calculable value of the vehicle dynamics. Examples are the yaw gain or the time difference between steering input and vehicle response (see also (Olley, 1947) (Rönitz, Braess, & Zomotor, 1977)). An OMC is defined by an analytical expression, which is a function of one or more FTC. If properly chosen an OMC correlates with one or more subjective manoeuvre indices (SMI) given by a test driver.

An appropriate set of OMC allows a unique description of the vehicle dynamics without using characteristic curves and without defining the vehicle's geometrical and material properties. It can be used to compare two vehicles or define the target dynamics of a vehicle.

3.1.4 Subjective Manoeuvre Index

A subjective manoeuvre index (SMI) defines a subjective value describing the vehicle dynamics as experienced by a test driver. Examples are the steering response or vehicle stability (see also (Heißing & Brandl, 2002)). A SMI cannot be defined by an analytical expression. If properly chosen it correlates with one or more OMC.

An appropriate set of SMI allows a unique description of the vehicle dynamics. It can be used to compare two vehicles or define the target dynamics of a vehicle.

3.2 Tyre Model

In this section, first, an overview of tyre models for vehicle simulation is given. Then, the choice of the Magic Formula Tyre Model and the MF-SWIFT Tyre Model are motivated.

3.2.1 Introduction

In the vehicle pre-development phase simulations are used to conduct parameter-variation studies and for identifying an optimal vehicle design. Simulations are necessary for reducing time and costs of the pre-development phase while increasing product quality and preventing conflicts. In the second half of the 20th century a large variety of tyre models is developed with different degrees of complexity. The necessity of different approaches followed the requirements of researchers and companies. In Figure 3.1 a possible classification of tyre models is presented (Pacejka, 2006).

Empirical models are based on mathematical equations with no physical meaning: these equations are fitted via regression techniques to test bench measurements. For the fitting, a very large number of full scale tests is needed: in fact, the tyre dynamics correctly reproduced by empirical models are limited to those measured on a test bench. Examples of empirical models are the Harty tyre model (Blundell & Harty, 2007), the Magic Formula model (MF model) introduced by (Bakker, Nyborg, & Pacejka, 1987) and the Magic Formula Short Wavelength Intermediate Frequency Tyre model (SWIFT model) based on the Ph.D. theses of (Zegelaar, 1998) and (Maurice, 2000). Similarity models are based on a limited number of equations needed to describe the basic tyre characteristics: these equations are fitted to test bench measurements. But, contrary to empirical models, similarity models can describe some off-nominal conditions through distortion, rescaling, and multiplication. The similarity approach is used first for vehicle dynamics by (Pacejka, 1958) and later implemented for tyre modelling by (Radt & Milliken, 1983).

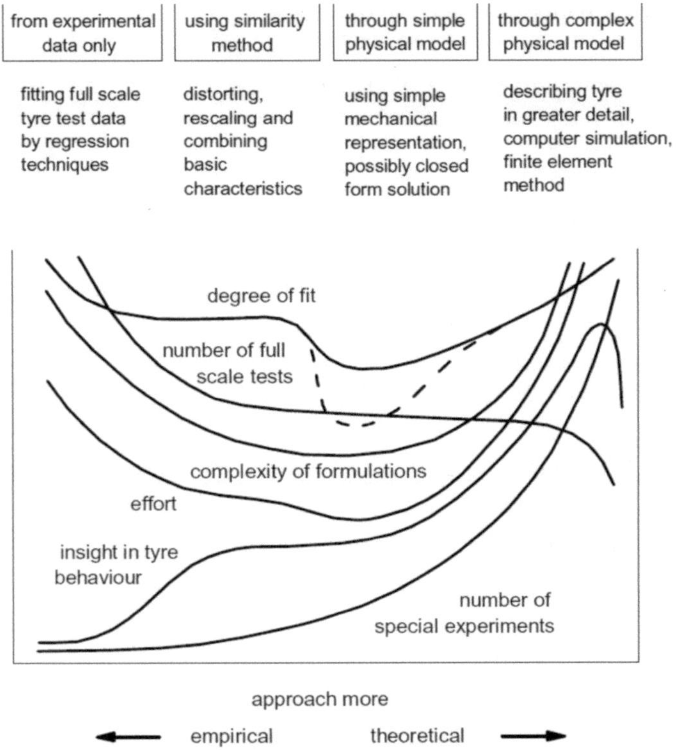

Figure 3.1: Different tyre models (Pacejka, 2006)

Simple physical models are based on physical equations needed to describe the most important physical tyre characteristics. Their formulation is simple, but provides sufficient accuracy for certain applications. A major advantage of physical tyre models is the possibility to simulate the influence on the tyre dynamics of tyre design parameters (e.g. the tyre width) without having measured it. Examples of physical models are the Brush Model, which is introduced by Fromm and Julien (cf. (Hadekel, 1952) for references) and then improved by Dugoff (Dugoff, Fancher, & Segel, 1970) and by Bernard (Bernard, Segel, & Wild, 1977). Further examples are the Brush and Ring Tyre model (BRIT) and the Comfort Tyre model (C-Tyre model) both introduced by (Gipser, 1998), the semi-physical Tyre Model easy (TMeasy) (Hirschberg, Rill, & Weinfurter, 2007) and the semi-physical tyre model Timis (Heinrichmöller, Benner, & Eckstein, 2014).

Complex physical models are based on an accurate description of the tyre geometrical and material properties. Typically they are based on finite element models (FEM), modal models or multibody models. They allow the description of flexible carcasses, non-steady-state behaviour and arbitrary pressure distribution and friction-coefficient variations. These models need special test bench measurements. Similarly to simple physical tyre models, they allow the investigation of tyre dynamics under operating conditions that have not been measured previously. Examples are the Dynamical Non-Linear Spatial Tyre model (DNS tyre model) introduced by (Gipser, 1987), the "Reifen Modell Komfort" model (RMOD-K model) (Oertel & Fandre, 2001) (Siebertz, van Bebber, & Hochkirchen, 2010), the Flexible Ring Tyre model (F-Tyre model) (Gipser, 2007) and the Comfort and Durability Tyre model (CD-Tyre) (Gallrein & Bäcker, 2007).

3.2.2 Choice of Tyre Models

In the present thesis, tyre models are used to analyse virtually the influence the FTCs of power loss, lateral dynamics and ride comfort have on the OMC. Instead, the analyses of the FTCs of interior noise are based on measurements. To obtain reliable results, large databases are needed. The tyre models should allow high accuracy and low computational effort. For the simulation of power loss and lateral dynamics the MF model 5.2 is chosen, as it represents a good compromise between accuracy and evaluation time. Moreover, a large database of tyre measurements and simulations is already available. Concerning ride comfort, the MF-SWIFT model 6.1.2 is chosen as it allows a good description of the tyre's vertical dynamics in the range of interest (up to 30 Hz) and as it is compatible with the MF model 5.2. In this section, the main characteristics of the two tyre models are presented.

3.2.2.1 MF model

The Magic Formula model (MF model) is a semi-empirical tyre model for reproducing tyre forces and torques under quasi-steady-state operating conditions. Typical applications are the simulation of the vehicle lateral dynamics.

The development of the Magic Formula starts in the 80's by a cooperation between the TU-Delft and Volvo: the sine and cosine-version of the Magic Formula are introduced in 1987 and are used for reproducing steady-state contact forces and torques of a rolling tyre (Niederreiter, 1988). In order to simulate also the tyre's transient behaviour, relaxation lengths are introduced (Bakker, Pacejka, & Lidner, 1989). In (Pacejka & Bakker, 1992) the combined slip calculation is improved and in the 90's the complexity of the model is reduced to allow faster simulations (Bayle, Forissier, & Lafon, 1993).

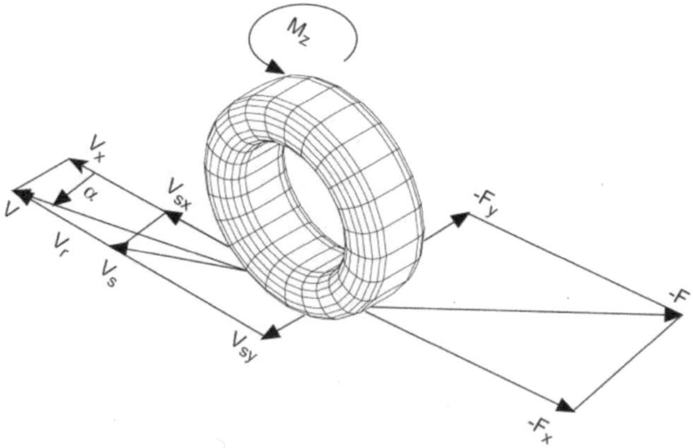

Figure 3.2: Forces and torques (MSC.Software Corporation, 2003)

In the present thesis, the version 5.2 of the MF model as implemented by (TNO Automotive, 2001) is used for all vehicle simulations. In this section, the most important equations of the MF model 5.2 are presented. For further details see (TNO Automotive, 2001).

As shown in Figure 3.2, the MF model 5.2 calculates longitudinal force F_x, lateral force F_y, overturning torque M_x, rolling resistance torque M_y and self-aligning torque M_z as a function of longitudinal slip κ, lateral slip α, wheel camber γ and vertical force F_z. Trajectory velocity v influences the effective rolling radius.

Pure longitudinal force $F_{x,0}$ and pure lateral force $F_{y,0}$ are modelled according to the general sine version of the Magic Formula (see Figure 3.3):

$$F_{i,0} = D_i \sin\{C_i \arctan[B_i\tilde{s} - E_i(B_i\tilde{s} - \arctan(B_i\tilde{s}))]\} + S_{v,i} \quad \textbf{3.1}$$

$$\tilde{s} = s + S_{h,i} \quad \textbf{3.2}$$

where $i \in [x, y]$, $s \in [\kappa, \alpha]$, B is the stiffness factor, C the shape factor, D the peak value, E the curvature, S_v the vertical shift and S_h the horizontal shift.

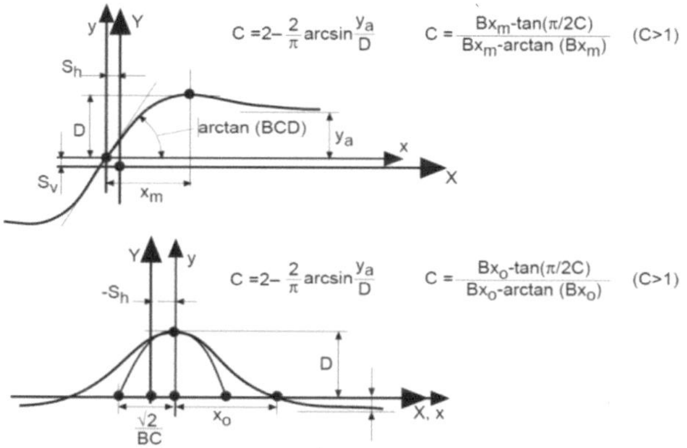

Figure 3.3: Sine and cosine versions of the Magic Formula (MSC.Software Corporation, 2003)

Pure self-aligning torque $M_{z,0}$ is a function of pure pneumatic trail t_0, pure lateral force $F_{y,0}$ and pure residual aligning torque $M_{z,r,0}$:

$$M_{z,0} = f(t_0, F_{y,0}, M_{z,r,0}) \quad \textbf{3.3}$$

Pure pneumatic trail t_0 is modelled according to the general cosine version of the Magic Formula (see Figure 3.3):

$$t_0 = D_t \cos\{C_t \arctan[B_t\tilde{\alpha}_t - E_t(B_t\tilde{\alpha}_t - \arctan(B_t\tilde{\alpha}_t))]\}\cos(\alpha) \quad \mathbf{3.4}$$

$$\tilde{\alpha}_t = \alpha + S_{h,t} \quad \mathbf{3.5}$$

The introduction of weighting-functions allows to model the combined effects of longitudinal and lateral slip. This affects both forces and the aligning torque:

$$F_x = F_{x,0} \cdot G_x(\alpha, \kappa, F_z) \quad \mathbf{3.6}$$

$$F_y = F_{y,0} \cdot G_y(\alpha, \kappa, \gamma, F_z) + S_{v2} \quad \mathbf{3.7}$$

$$M_z = f(t, F_x, F_y, M_{z,r}) \quad \mathbf{3.8}$$

where G_i are the weighting functions and S_{v2} an additional vertical shift for the combined lateral force. It should be noticed, that the longitudinal force is not dependent from camber angle. For modelling the transient behaviour of longitudinal force F_x and lateral force F_y up to frequencies of $8\ Hz$ a set of differential equations based on longitudinal relaxation length σ_x and lateral relaxation length σ_y is used. Finally, the overturning torque M_x is a function of:

$$M_x = f\left(\gamma, F_z, F_y(\alpha, \kappa, \gamma, F_z)\right) \quad \mathbf{3.9}$$

whereas rolling resistance torque M_y is a function of:

$$M_y = f(F_z, v_x, F_x(\alpha, \kappa, F_z)) + M_{y,0} \quad \mathbf{3.10}$$

where v_x is the longitudinal velocity of the tyre. It should be noticed, that the variation of rolling resistance torque is mainly dependent from tyre load and trajectory velocity.

3.2.2.2 MF-SWIFT model

The Magic Formula Short Wavelength Intermediate Frequency Tyre model (MF-SWIFT model) is a semi-empirical tyre model that combines the slip force description of the MF model with a rigid ring model. The MF-SWIFT model is valid up to frequencies of 60-100 Hz and for short wavelengths ($< 0.2\,m$) (Schmeitz, Besselink, & Jansen, 2007). Typical applications are the simulation of ride comfort, as well as durability studies, shimmy analysis and chassis control system evaluation.

In the 1990s the Delft University and the TNO start a joint research project with the aim to develop a tyre model that can be used for handling manoeuvres on uneven road in order to develop active control systems like anti-lock brake system (ABS) traction control system (ASR, TCS) and active yaw control systems (ESP, VDC). The first version of the in-plane and out-of-plane model are introduced by (Zegelaar, 1998) and (Maurice, 2000). Afterwards, the focus of the model changed to ride comfort and durability studies (Schmeitz, 2004). A comprehensive description of the last developments can be found in (Pacejka, 2006).

In the present thesis the version 6.1.2 of the MF-SWIFT model as implemented by (TNO Automotive, 2010) is used for all vehicle simulations. In this section, the most important equations of the MF-SWIFT 6.1.2 are presented. For further details see (TNO Automotive, 2010).

The MF-SWIFT model consists of four elements:

- Magic Formula
- Contact patch slip model
- Rigid ring
- Obstacle enveloping model

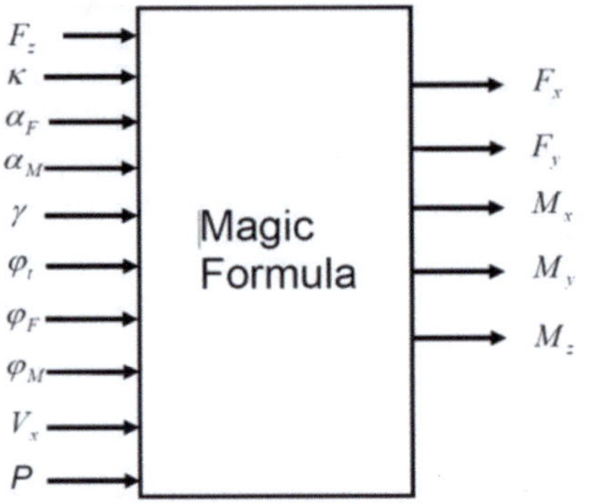

Figure 3.4: Magic Formula for the MF-SWIFT model 6.1.2 (TNO Automotive, 2010)

The equations of the Magic Formula are identical to those presented in section 3.2.2.1, but include also an input variable for turn slip φ_t and spin φ_F. In Figure 3.4 inputs and outputs are represented schematically. It should be noted, that for steady state conditions the values of slip angle and turn spin for lateral force calculation (α_F, φ_F) are equal to those for self-aligning torque calculation (α_M, φ_M); this is not true for transient behaviour.

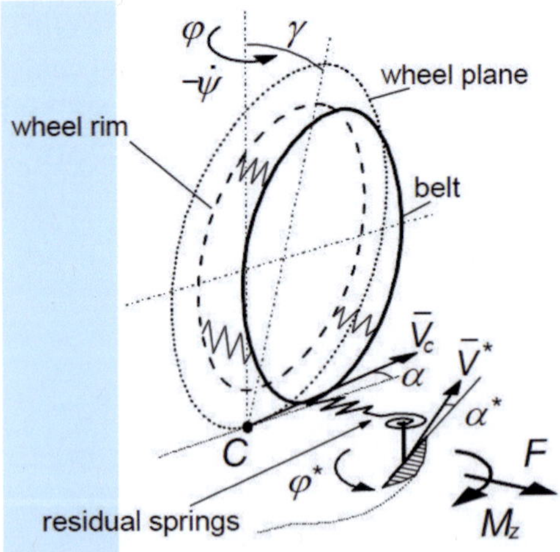

Figure 3.5: Contact patch and rigid ring model of the MF-SWIFT model 6.1.2 (TNO Automotive, 2010)

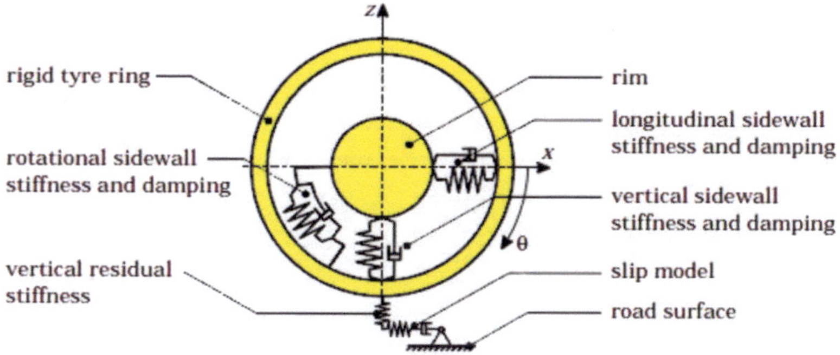

Figure 3.6: Rigid ring and contact patch slip model (Zegelaar, 1998)

The contact patch slip model consists of differential equations modelling the relaxation behaviour of the contact patch. The contact patch is elastically connected through residual springs to the belt (see Figure 3.5): the forces at the contact patch act on small masses and are then transferred to belt and wheel hub. Subsequently, the slip angle of the contact patch differs from the slip angle of the rim: it is a function of relaxation length and carcass stiffness.

Scientific researches show (Zegelaar, 1998) (Maurice, 2000), that deformations of the tyre belt can be neglected for modelling the tyre dynamic behaviour up to frequencies of 60-100Hz, but it is no longer allowed to consider rim and belt as one rigid entity. In Figure 3.5 the six degrees of freedom of the belt with respect to contact patch and rim are shown. In Figure 3.6 a representation of the only in-plane movements of the belt are represented, as introduced by (Zegelaar, 1998).

In order to model short wavelengths ($< 0.2\ m$) an enveloping model is needed. In (Schmeitz, 2004) a technique using elliptical cams is developed to transform the real road surface in the effective road surface (see Figure 3.7): elliptical cams sense the road undulations allowing to calculate the effective road surface for arbitrary 3D road unevenness.

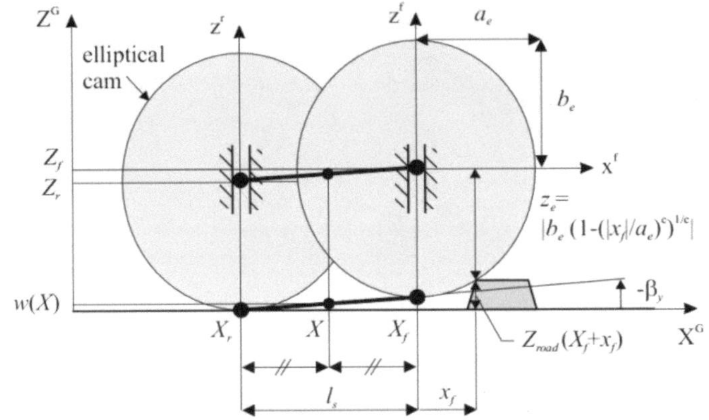

Figure 3.7: Contact patch model of the MF-SWIFT model 6.1.2 (TNO Automotive, 2010)

3.3 Generation of Virtual Tyres

In this section, the Tyre Shaper and the Tyre Fitter are presented: these tools are based on a genetic algorithm and allow respectively the generation of artificial tyres and the parameterization of an MF-tyre.

3.3.1 Tyre Shaper

The Tyre Shaper is introduced by (Niedermeier, Peckelsen, & Gauterin, 2013); further developments concerning rolling resistance are presented in (Peckelsen, Gauterin, & Unrau, 2015). Its implementation in the tyre development process is described in (Niedermeier, 2015). The Tyre Shaper is a tool used to generate MF-parameter-sets that have desired physical characteristics. To this purpose it optimises the MF-parameters so, that the characteristic curves of the Magic Formula approximate best the desired FTCs.

The structure of the genetic algorithm is based on the classical genetic algorithm theory (Davis, 1991) and Darwin's Theory of Evolution. Figure 3.8 represents a typical evolution cycle of a genetic algorithm.

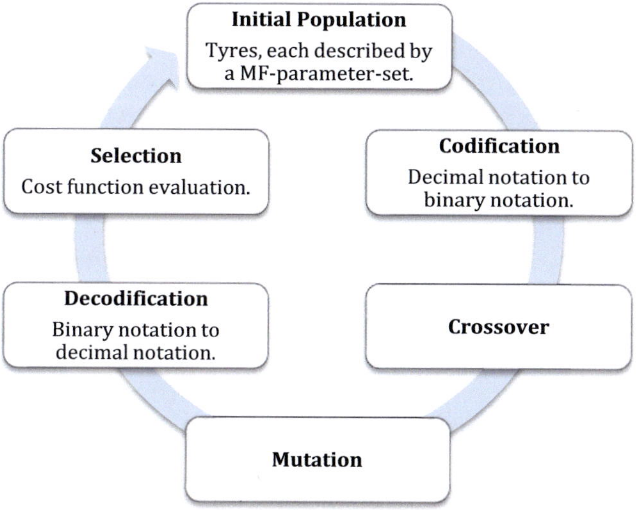

Figure 3.8: Genetic algorithm cycle

During each cycle new individuals (children), here MF-parameter-sets, are generated through crossover and mutation of the genes of previous individuals (parents). In order to perform crossover and mutation, genes of each individual are codified into binary notation. Before selection, the fitness of each individual is evaluated: it is measured according to the cost function (see 3.16). During selection, only the individuals with the highest fitness, i.e. lowest fitting error, survive.

The cost function of the Tyre Shaper is:

$$Err = \sum_{i=1}^{n} \frac{|Y_i - \hat{Y}_i|}{k_i} \cdot 100 \qquad \textbf{3.11}$$

where Y_i and $\hat{Y}_i$ are the values of the i-th FTC of fitted and desired tyre, and k_i is a weight factor defined as the difference between possible maximum ($Y_{i,max}$) and possible minimum ($Y_{i,min}$) FTC value:

$$k_i = Y_{i,max} - Y_{i,min} \qquad \textbf{3.12}$$

This cost function is defined symmetric, continuous and takes into account the order of magnitude and, thanks to the introduced weighting factors, the domain of definition of the optimised quantities (Peckelsen, 2012). A global optimization is guaranteed, as the starting population is created through a quasi-random sequence (no risk of local optima due to a fixed starting point) and as the generation process (crossover and mutation) is performed randomly (no risk of local optima). Convergence of optimization is guaranteed, as at each cycle only the genes of the best individuals are used to generate new individuals: the selection process fosters the generation of individuals with low fitting error.

A multi-crossover strategy and a decreasing function for population and for mutation probability are implemented in order to improve the efficiency of the optimization, i.e. maximize accuracy and minimize time:

$$pop(t+1) = (pop_0 - pop_{min})\left(\frac{Err(t)}{Err(\bar{t})}\right)^{\alpha} + pop_{min} \qquad \textbf{3.13}$$

where pop_0 is the starting population of individuals, i.e. tyres, pop_{min} the minimum population and α the decrease factor. In contrast to (Peckelsen, 2012), these functions are adopted only after a given number $\bar{t}$ of optimisation cycles. The position where the gene is cut for crossover is defined by the integer i_{cross}:

$$i_{cross} = round\left(x^s \cdot (n_{bit} - 1)\right) \qquad \textbf{3.14}$$

where x is a random value between 0 and 1, n_{bit} is the number of bits used for codifying each gene and s is the crossover shape function:

$$s(t+1) = (s_{max} - s_0)\left(1 - \frac{Err(t)}{Err(\bar{t})}\right)^{\gamma} + s_0 \qquad \textbf{3.15}$$

where $s_0 = 1$ is the starting shape value, s_{max} the maximum shape value, $Err(t)$ the actual fitting error, $Err(\bar{t})$ the fitting error after a given number $\bar{t}$ of cycles and γ a factor. High values of the shape function increase the probability of choosing small i_{cross} values. Subsequently, if the actual solution is close to the target value ($Err = 0$), only changes in decimal places of the genes are allowed, improving the accuracy of the optimization.

3.3.2 Tyre Fitter

The Tyre Fitter (Peckelsen, Gauterin, & Unrau, 2015) allows the parameterization of MF-tyres, i.e. it generates MF-parameter-sets that fit best measurements performed either on a FlatTrac or on a test drum bench. It is developed in order to adapt easily the parameterization of MF-tyres to new measurement routines, e.g. the rolling resistance measurement routine presented in section 4.1.2. Moreover, it allows a global fitting, which is independent from the starting MF-parameter-set. A similar approach is implemented by (Cabrera, Ortiz, Carabias, & Simon, 2004). The structure of the genetic algorithm is identical to those of the Tyre Shaper. The cost function of the Tyre Fitter is:

$$Err = i{=}1nj{=}1mXi,j{-}Xi,jwi\cdot 100 \qquad \textbf{3.16}$$

where X_i and $\hat{X}_{i,}$ are vectors of length m containing respectively the values of the i-th modelled and measured quantity X_i (e.g. lateral force over slip angle), and w_i is a weight factor defined as the difference between possible maximum and minimum value of X_i:

$$w_i = X_{i,max} - X_{i,min} \qquad \textbf{3.17}$$

All considerations made for the Tyre Shaper concerning the cost function and crossover strategy are valid for the Tyre Fitter, too.

3.4 Design of Experiment

In this section, first, a brief introduction and overview of the most used design of experiments is given; then, the quality of the design is defined and the global homogeneity index, a new index for measuring the homogeneity of a design, is introduced.

3.4.1 Introduction

Designs of Experiments (DoE) are necessary for numerical integrations, optimisation problems and correlation analyses. To these applications the quality of the DoE is crucial: in fact, depending on the problem under consideration, it may affect to a major or minor extend the effectiveness and reliability of the result.

A DoE is characterized by:

- n_{DV} design variables, which are the input parameters of the examined system;
- n design experiments, defined by combinations of the design variables (n is the sample size);
- the design generation method, which is needed to determine the values of the design variables for each design experiment;
- n_{DS} design solution, which represent the output values of the examined system.

A DoE has a design variable space, given by the domains of the design variables, and a design solution space, given by the domains of the design solutions.

In this section, a classification of the most important design generation methods is presented.

3.4.1.1 Factorial Designs

Full-factorial designs (FFD) consist of all combinations of the k subintervals of the n_{DV} design variables (Siebertz, van Bebber, & Hochkirchen, 2010). The experiments

$$n = k^{n_{DV}} \quad \textbf{3.18}$$

are positioned deterministically on a "grid". The characteristic of a full-factorial design is a completely homogeneous distribution of design experiments in the design variable space. This improves the effectiveness of numerical integrations and optimisation algorithms, and the accuracy of sensitivity analyses in identifying main and side effects. However, for most applications the time needed to evaluate the experiments is impracticable, as the sample size growths exponentially with the number of design variables. Moreover, concerning sensitivity analysis, the contribution of side effects is often very small removing the need to study all of them (Kleppmann, 2008).

Fractional-factorial designs (FrFD) consist of subsets of full-factorial designs. The selection of the subsets causes a superposition of main and side effects of the design variables. This superposition is known as "confounding" (Kleppmann, 2008). Compared to full-factorial designs, fractional-factorial designs have the advantage of reducing the number of experiments to:

$$n = k^{n_{DV}-n_C} \quad \mathbf{3.19}$$

where n_C represents the number of design variables whose effects are confounded. The selection of the subsets should avoid the confounding of main effects and low-order-interactions: the resolution (see Table 3.9) should be higher than III (Siebertz, van Bebber, & Hochkirchen, 2010). Fractional-factorial designs with a resolution higher than V are used seldom, as their evaluation-time would be comparable to that of full-factorial designs.

Table 3.9: Resolution of fractional-factorial designs

Resolution	Confounding
II	Main effects and side effects are confounded.
III	Main effects are not confounded by main effects. Side effects of second and higher order are confounded.
IV	Main effects are not confounded by side effects of second order. Side effects of third and higher order are confounded.
X	Main effects are not confounded by side effects of "X-2" order. Side effects of "X-1" and higher order are confounded.

3.4.1.2 (Quasi-) Random Sequences

Random sequences (RS) consist of experiments that are distributed randomly in the design variable space: the sample size can be arbitrarily chosen. RSs are implemented easily, but the haphazardness of the method affects its quality and, subsequently, the reliability of the result (Siebertz, van Bebber, & Hochkirchen, 2010).

Low discrepancy sequences (LDS), a subcategory of quasi-random sequences, consist of experiments that are distributed deterministically in the design variable space according to an analytical formula: the sample size can be arbitrarily chosen. The formulation of LDS is based on the definition of discrepancy. The discrepancy D measures the non-uniformity of n samples $\boldsymbol{x}_1, \dots, \boldsymbol{x}_n$ placed in a half-open d-dimensional unitary hypercube $I^d = [0\ 1)^d$ (Niederreiter, 1988):

$$D = \sup_{\forall J} |A(J;\ n) - V(J) \cdot n| \qquad \textbf{3.20}$$

where $A(J;\ n)$ is the number of samples that can be found in J, and $V(J)$ is the volume of the half-open subinterval J. The discrepancy D would be equal to zero for any half-open subintervals $J \in I^d$, if all samples are uniformly distributed in the design variable space. The experiments $\boldsymbol{x}_1, \dots, \boldsymbol{x}_n$ generate a LDS, if:

$$D \leq \mathrm{c}(\log(n))^d + O((\log(n))^{d-1}) \qquad \forall n > 1 \qquad \textbf{3.21}$$

The constant c, which depends on the design-variable-space dimension d, should be as small as possible. Currently, it is known that Niederreiter's formulations have the lowest c.

In bibliography, the most famous LDSs are, in chronological order: Halton's (Halton, 1960), Sobol's (Sobol, 1967), Faure's (Faure, 1982) and Niederreiter's sequence (Niederreiter, 1992). A similar method to generate LDS is the well-known Latin Hypercube (McKay, Beckman, & Conover,

1979). Furthermore, permutation algorithms try to improve the homogeneity of LDSs. An example is Matousek's scramble for Sobol's LDS (Matousek, 1998).

3.4.1.3 Maximin Designs

Maximin designs consist of experiments that are distributed in the design variable space according to an optimisation algorithm, the sample size can be arbitrarily chosen. Maximin designs are based on the assumption that that a design is good, if the samples are placed far from each other. Let $X = (x_{ij})$ be the $n \ x \ n_{DV}$ design matrix that contains on each row the experiment $i = 1 \dots n$ defined by all design variables $j = 1 \dots n_{DV}$. The Euclidean distance between two experiments is:

$$d_{ij} = \left\|\boldsymbol{x}_i - \boldsymbol{x}_j\right\|_2 = \sqrt{\sum\nolimits_k \left(x_{i,k} - x_{j,k}\right)^2} \qquad 3.22$$

The design matrix X^* of all possible design matrices X is a maximin design, if and only if $X^* \in E$ is a global solution of the optimisation problem (Johnson, Moore, & Ylvisaker, 1990), where E denotes the set of sites:

$$\begin{array}{ll} maximise & min(d_{ij}) \\ subjected\ to & X^* \in E \end{array} \qquad 3.23$$

Maximin designs are very appealing, as they allow optimal distributions. Moreover, their definition does not depend on the geometry of the design space; subsequently, they are used often for designs having non-rectangular design variable spaces (Trosset, 1999). However, as an optimisation problem has to be solved, the time required for generating the DoE is much longer if compared to quasi-random sequences[2].

3.4.2 Quality of a DoE

The quality of a DoE is determined by the choice of the design variables, of the sample size and of the generation method. The present section focuses on the choice of the generation method; the definition of the design variables and the sample size are assumed to be given.

A possible criterion for measuring the non-uniformity of a sequence is the discrepancy D (see section 3.4.1.2):

$$D = \sup_{\forall J} |A(J;\, n) - V(J) \cdot n| \tag{3.24}$$

where $A(J;\, n)$ is the number of samples x_n that can be found in J, and $V(J)$ is the volume of the half-open subinterval J. However, the computational effort needed to compute the discrepancy increases exponentially with dimension, as all subintervals $J \in I^d$ has to be analysed, where I^d is a half-open d-dimensional unitary hypercube $I^d = [0,1)^d$. Moreover, discrepancy is influenced by size and position of the subintervals J. Therefore, the star discrepancy D^* is introduced:

$$D^* = \sup_{\forall J \in E^*} |A(J;\, n) - V(J) \cdot n| \tag{3.25}$$

[2] Measurements of the evaluation-time required to generate LDS can be found in (Morokoff & Caflisch, 1994); to solve the optimisation problem of a Maximin Design in (Trosset, 1999). E.g. Maximin Design takes 0.65 s for 16 samples in two dimensions (it grows rapidly by increasing the number of samples and dimensions); whereas Sobol's or Halton's sequence take just 0.02 seconds for 1000 points in five dimensions.

where E^* is the set of sub-rectangles with one corner at zero. So, the sub-intervals J are limited in number and have a fixed size and position. However, this formulation represents an approximation of the discrepancy D, decreasing accuracy in predicting the non-uniformity of the sample distribution.

In order to avoid an exponentially increasing computational effort and an approximation for calculating the quality of the design, a new index is introduced (Peckelsen, 2012): the global homogeneity index. It is based on the assumption that a design is good, if the samples are placed far from each other (Johnson, Moore, & Ylvisaker, 1990). The global homogeneity index consists of three sub-indices: the first two indices measure respectively the homogeneity of the design and the mean homogeneity along single dimensions; the third index measures the correlation between the design variables. In section 3.4.2.1, section 3.4.2.2 and section 3.4.2.3 the sub-indices of the global homogeneity index for measuring the quality of designs are presented. In section 3.4.3, well-known designs are compared for different combinations of dimension and sample size.

3.4.2.1 Homogeneity of the Design

In this section, the first sub-index of the global homogeneity index is presented: it measures the homogeneity of the whole design. Let $M = [\boldsymbol{x}_i]$ be the $n \; x \; d$ design matrix containing n samples $\boldsymbol{x}_1, \dots, \boldsymbol{x}_n$ placed in a half-open d-dimensional unitary hypercube $I^d = [0,1)^d$. Each sample $\boldsymbol{x}_i$ is a vector with d elements:

$$\boldsymbol{x}_i = \{x_{i,1}, \dots, x_{i,d}\} \qquad \textbf{3.26}$$

The Euclidean distance d_{ij} between two samples in d dimensions is:

$$d_{ij} = \left\|\boldsymbol{x}_i - \boldsymbol{x}_j\right\|_2 = \sqrt{\sum_{k=1}^{d} \left(x_{i,k} - x_{j,k}\right)^2} \qquad \textbf{3.27}$$

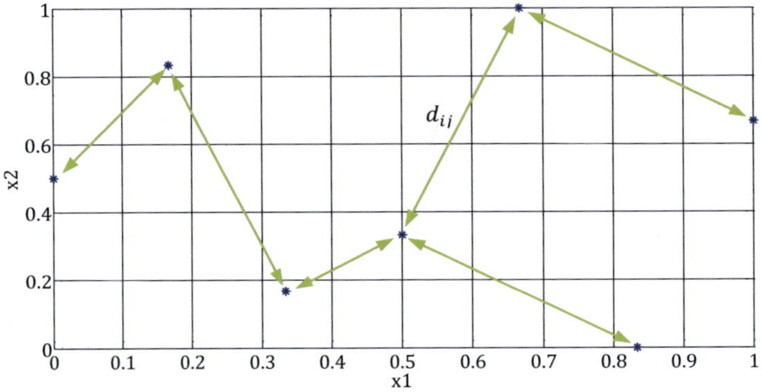

Figure 3.10: Example of a net for two dimensions and seven samples

For each design it is possible to define a net represented by the vector $\boldsymbol{d}_M$ containing the $n-1$ shortest distances of all possible distances d_{ij}. These distances connect all samples so, that there exist no other net $\boldsymbol{d}_N$ whose sum of minimum distances is smaller:

$$\sum_{n-1} \boldsymbol{d}_M \leq \sum_{n-1} \boldsymbol{d}_N \qquad \textbf{3.28}$$

An example of a net for two dimensions and seven samples is presented in Figure 3.10. For each combination of dimension and sample size a design $R = [\boldsymbol{x}_i]$ exists, whose minimum distances in vector $\boldsymbol{d}_R$ are equal and whose sum of distances is bigger than that of any other net $\boldsymbol{d}_N$:

$$\sum_{n-1} \boldsymbol{d}_R \geq \sum_{n-1} \boldsymbol{d}_N \qquad \textbf{3.29}$$

Let us take full-factorial designs as reference designs, as they have the highest homogeneity for their combination of dimension and sample size. Let $R = [\boldsymbol{x}_i]$ be the design matrix of a full-factorial design. Then the minimum distance between its samples is:

$$d_{ref} = \frac{1}{n^{1/d} - 1} \qquad \textbf{3.30}$$

and is here defined as the reference distance d_{ref}. For all combinations of dimension and sample size that are not full-factorial designs, the reference distance is approximated interpolating logarithmically the reference distances d_{ref} of the full-factorial designs (Peckelsen, 2012).

The homogeneity of the design is ($H^{doe} \in [0,1]$):

$$H^{doe} = 1 - \sum_{n-1} \frac{|d_{ref} - \boldsymbol{d}_M|}{s} \qquad \textbf{3.31}$$

where s is a scaling factor defined as the difference between reference distance d_{ref} and the distances $\boldsymbol{d}_W$ of the design with the lowest homogeneity in d dimensions:

$$s = \sum_{n-1} |d_{ref} - \boldsymbol{d}_W| \qquad \textbf{3.32}$$

The design with the lowest homogeneity in d dimensions has $n-1$ sample located in one corner, and one sample located in the opposite corner. Subsequently, the net $\boldsymbol{d}_W$ contains $n-2$ distances of length 0, and one distance of length $\sqrt{d}$:

$$\boldsymbol{d}_W = \{0, \dots, 0, \sqrt{d}\} \qquad \textbf{3.33}$$

The expression of scaling factor s is then:

$$s = (n-3)d_{ref} + \sqrt{d} \qquad \textbf{3.34}$$

3.4.2.2 Mean Homogeneity along Single Dimensions

In this section, the second sub-index of the global homogeneity index is presented: it measures the homogeneity along single dimensions.

Similarly to the first index, the mean homogeneity along all single dimensions d is ($H^* \in [0,1]$):

$$H^* = 1 - \frac{1}{d}\sum_{\bar{k}=1}^{d}\sum_{n-1}\frac{\left|d_{ref}^* - \boldsymbol{d}_{M,\bar{k}}^*\right|}{s^*} \qquad 3.35$$

The net $\boldsymbol{d}_{M,\bar{k}}^*$ contains the $n-1$ minimum distances of all possible distances $d_{ij,\bar{k}}^*$ along a single dimension $\bar{k} = 1 \ldots d$:

$$d_{ij,\bar{k}}^* = \left\|x_{i,\bar{k}} - x_{j,\bar{k}}\right\|_2 \qquad 3.36$$

Reference distance d_{ref}^* and scaling factor s^* are calculated according to equation 3.30 and equation 3.34 substituting $d = 1$.

$$d_{ref}^* = \frac{1}{n-1} \qquad 3.37$$

$$s^* = (n-3) \cdot d_{ref}^* + 1 \qquad 3.38$$

For this sub-index, no approximation is needed to calculate the reference distance.

3.4.2.3 Mean Correlation Index

In this section, the third sub-index of the global homogeneity index is presented: it measures the correlation between two variables.

To compute the mean correlation index, Spearman's index is chosen as it is more robust than Pearson's index especially if samples are not normally distributed. The correlation between two dimensions k and l according to Spearman measures the monotony of the relation between two variables (Gobbi, Mastinu, & Miano, 2005):

$$\rho_{S,kl} = \frac{cov\left(R\left(\boldsymbol{x}_{1:n,k}\right), R\left(\boldsymbol{x}_{1:n,l}\right)\right)}{\sigma_{R(\boldsymbol{x}_{1:nk})}\,\sigma_{R(\boldsymbol{x}_{1:n,l})}} \qquad 3.39$$

where cov is the covariance, σ the standard deviation and R the rank (see section 3.5.1.2). The mean correlation index between all combinations of dimension k and l is ($H^{corr} \in [0,1]$):

$$H^{corr} = \frac{2}{d(d-1)} \sum\nolimits_{\forall k, \forall l>k} |\rho_{S,kl}| \qquad \textbf{3.40}$$

3.4.2.4 Global Homogeneity Index

The global homogeneity index H is a weighted sum of all three indices:

$$H = w_1 H^{doe} + w_2 H^* + w_3(1 - H^{corr}) \qquad \textbf{3.41}$$

where the weights w can be chosen by the user. In our case the weights are set according to the following considerations. The homogeneity index H^{doe} is the most important index as it determines directly the homogeneity of the design. Its weight is set to $w_1 = 0.70$. Low correlation is always important as it assures that two variables can be analysed separately: its weight is set to $w_3 = 0.25$. The homogeneity index H^* is crucial only for few samples ($n < 5$): in this case, only a homogeneous distribution allows to detect non-linearities of the relation between design solutions and design variables. In general, also low values of the index H^* are sufficient. Therefore, it is weighted less than the other two indices ($w_2 = 0.05$).

The so defined homogeneity index allows a fast, accurate and unique measurement of the quality of the design:

- homogeneity index H^{doe} and correlation index H^{corr} measure the informative value of the design (i.e. correct identification of main and combined effects for analyses; high convergence and high probability of finding the global optimum for evolutionary algorithms).
- homogeneity index H^* measures the accuracy of identifying non-linearities of the main relations between design variables and design solutions.

In contrast to the definition of discrepancy, the global homogeneity index is defined by an average value that takes into account the relative position of all samples of the distribution: the index is neither dependent on the position and the size of sub-intervals nor it focuses only on the worst case.

Moreover, the computational effort needed to calculate the global homogeneity index depends only on the number of experiments; it does not increase exponentially with dimension.

3.4.3 Comparison between Different Designs

In literature comparisons between different designs referring to the definition of discrepancy are made, e.g. (Morokoff & Caflisch, 1994). In this section a comparison based on the introduced global homogeneity index is presented. First, a brief description of selected designs is given. For generating them standard Matlab functions (R2010b V 7.11.1.886) are used.

Random Sequences (RS) are easily implemented and generate random distributions rapidly. Generally, samples of RS are not well-distributed, as the haphazardness of the method affects its quality (Siebertz, van Bebber, & Hochkirchen, 2010). Low Discrepancy Sequences (LDS), a particular type of quasi-random sequences, are based on mathematical equations that deterministically generate a distribution for a given dimension and sample size. Well-known examples are, in chronological order: Halton's Sequence (HS) (Halton, 1960), Sobol's Sequence (SS) (Sobol, 1967), Faure's Sequence (Faure, 1982) and Niederreiter's Sequence (Niederreiter, 1992). They perform rapidly and generate homogeneous designs, if dimension and sample size are not too high (Morokoff & Caflisch, 1994). Latin Hypercube Sampling (LHS) (McKay, Beckman, & Conover, 1979) represent an efficient approach to generate quasi-random sequences: the design space is subdivided in equally-sized subintervals, each occupied by at most one sample. The algorithm that chooses which subinterval should be occupied can minimise correlation (LHS_c) or maximise the minimum distance (LHS_m) between the samples. LDS can use permutation algorithms (e.g. Sobol Sequence with Matousek Scramble SSS (Matousek, 1998)): these algorithms try to improve the quality of the distribution for those combinations of dimension and sample size for which the sequence (deterministically) generates poor designs.

3.4.3.1 Different Designs in two Dimensions

In this section, a comparison between designs in two dimensions is presented. This case study allows to compare the numerical evaluation performed by the global homogeneity index to well-known characteristics of the design (e.g. poor homogeneity, high correlation).

In Figure 3.11 the result for a RS is presented. It is clearly visible, how some samples are closely spaced (red circle), leaving some areas completely unexplored (blue dotted circle). This characteristic is reflected by low values of the homogeneity indices H^{doe} and H^*.

In Figure 3.12 and Figure 3.13 two designs generated by LHS are presented. The first algorithm minimises correlation, the second maximises the minimum distance between the samples. The indices reflect this: correlation index H^{corr} is lower for the first design, homogeneity index H^{doe} is higher for the second one. This also proves the necessity of introducing two indices, the homogeneity index and the correlation index: there is no one-to-one relation between maximizing minimum distances (i.e. homogeneity) and minimizing correlation, as already observed by (Joseph & Hung, 2008). Furthermore, it should be noticed that the homogeneity index H^* is exactly one for the first design: in fact, if the position of the samples of a Latin Hypercube is not varied in order to maximises the minimum distance, they are placed equidistantly along each dimension.

Figure 3.14 and Figure 3.15 show the results for Sobol's and Halton's LDS. Although correlation is poor, both sequences give good results. Especially the homogeneity of the distribution of HS is high: in fact, HS are known to generate high quality distributions with few samples. On the contrary, with increasing number of samples the quality of SS outperforms that of HS.

The last two figures, Figure 3.16 and Figure 3.17, show the results of SS and HS using permutation (scramble) algorithms. In this case homogeneity does not improve significantly, for HSS it also decreases, but correlation is better (i.e. lower): SSS benefits from the permutation, HSS not.

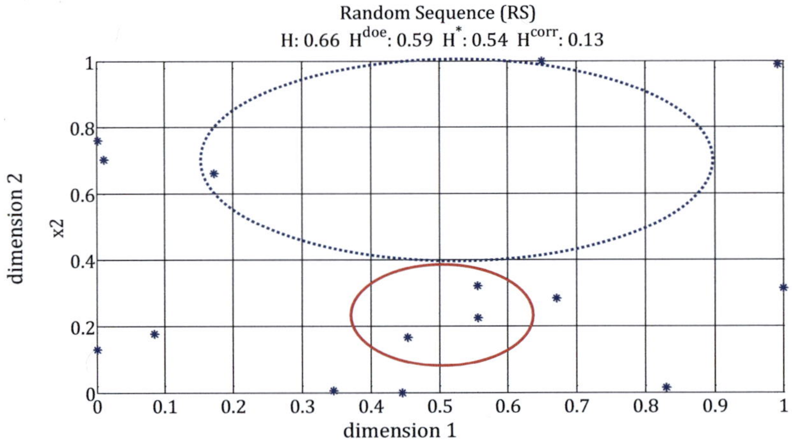

Figure 3.11: Comparison in two dimensions (RS)

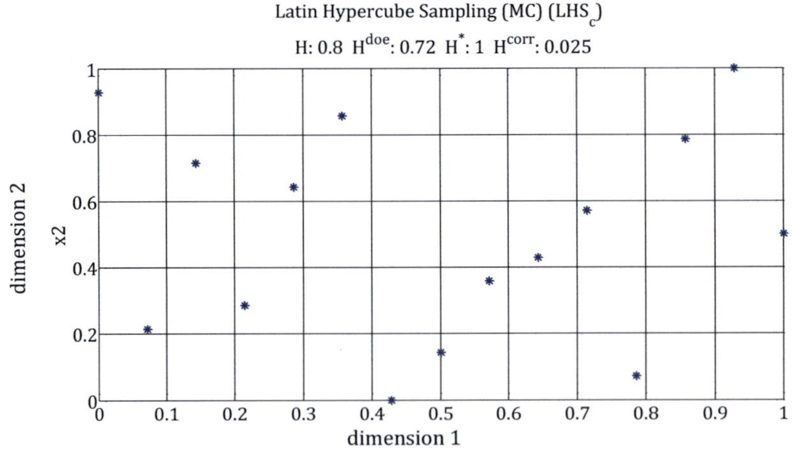

Figure 3.12: Comparison in two dimensions (LHS correlation)

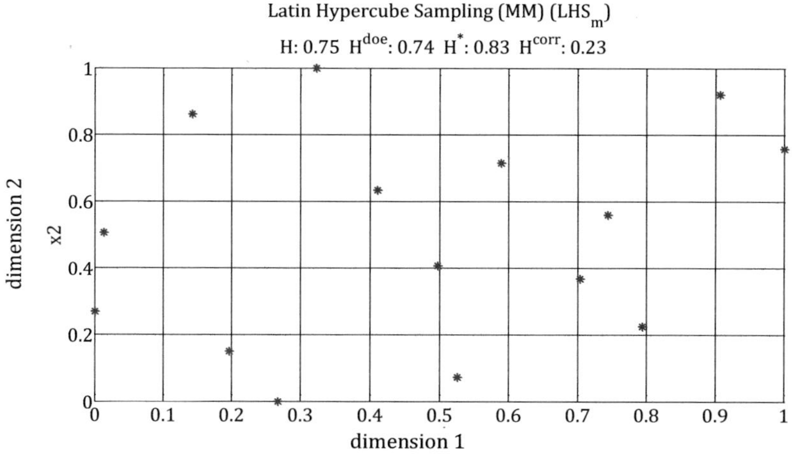

Figure 3.13: Comparison in two dimensions (LHS minimax)

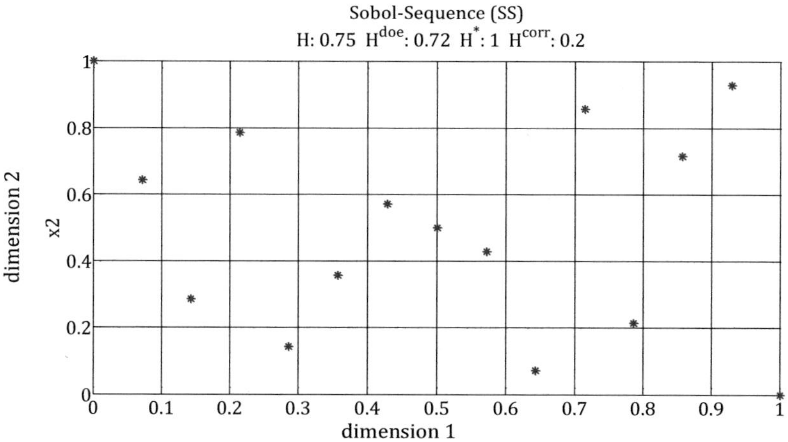

Figure 3.14: Comparison in two dimensions (SS)

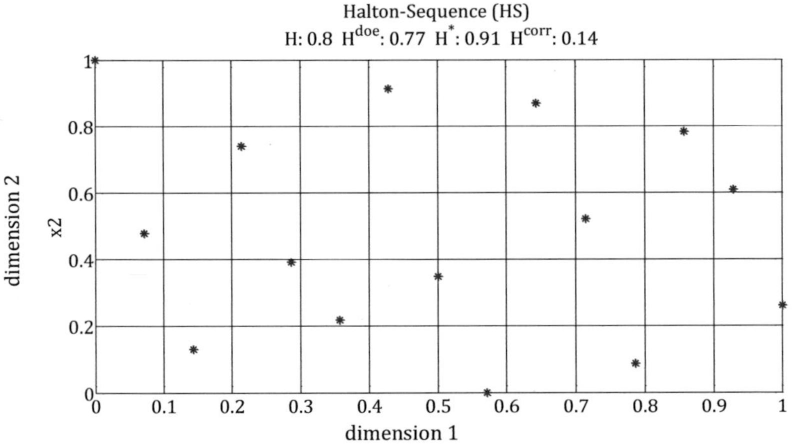

Figure 3.15: Comparison in two dimensions (HS)

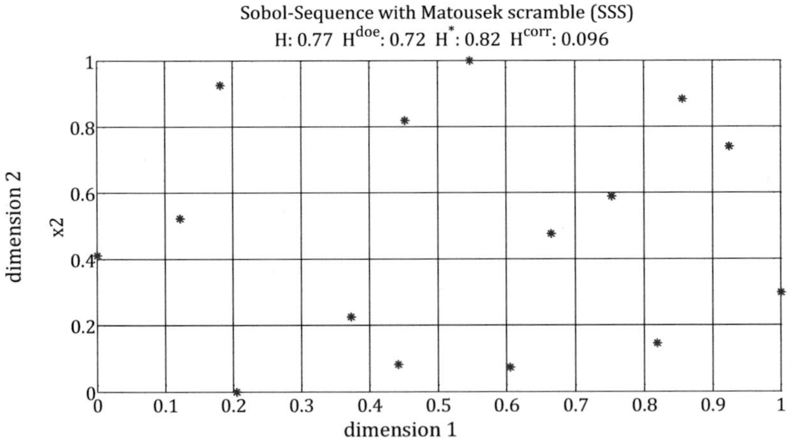

Figure 3.16: Comparison in two dimensions (SSS)

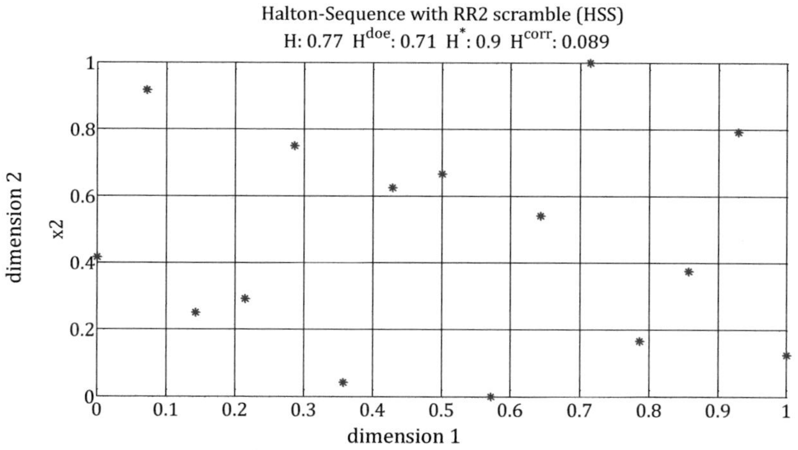

Figure 3.17: Comparison in two dimensions (HSS)

3.4.3.2 Different Designs in n Dimensions

In this section, a graphical comparison of selected designs is given: the homogeneity of the distribution is calculated as a function of both dimension and sample size. The course of homogeneity as a function of the sample size is presented for 7, 15 and 50 dimensions (see Figure 3.18, Figure 3.19 and Figure 3.20).

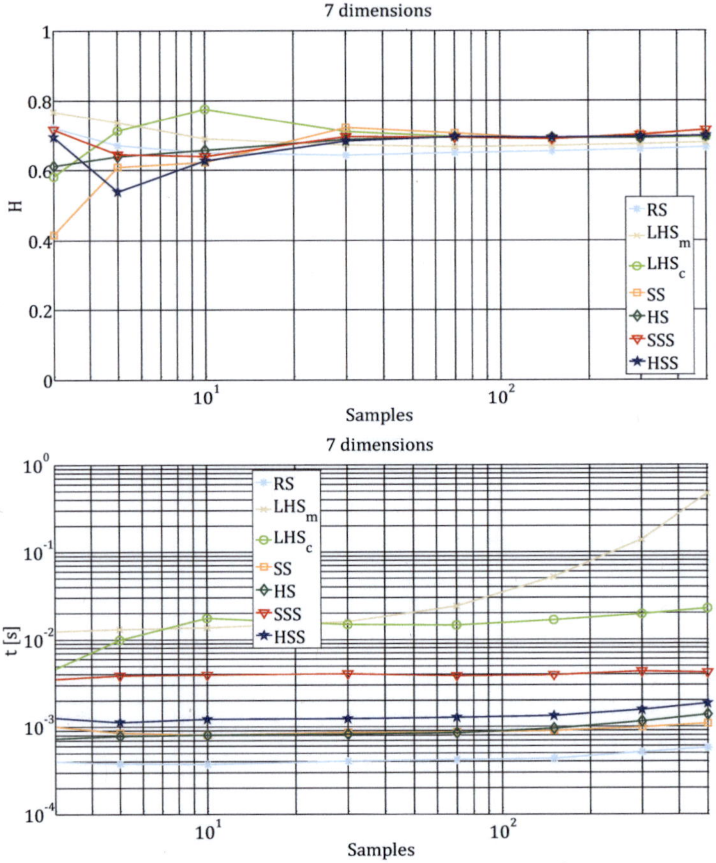

Figure 3.18: Comparison in 7 dimensions

The values of the homogeneity indices for the different DoEs is an average based on 50 repetition. Moreover, generation time t is calculated, although it is often negligible if compared to the time necessary to collect the data for the analysis or to perform the optimisation.

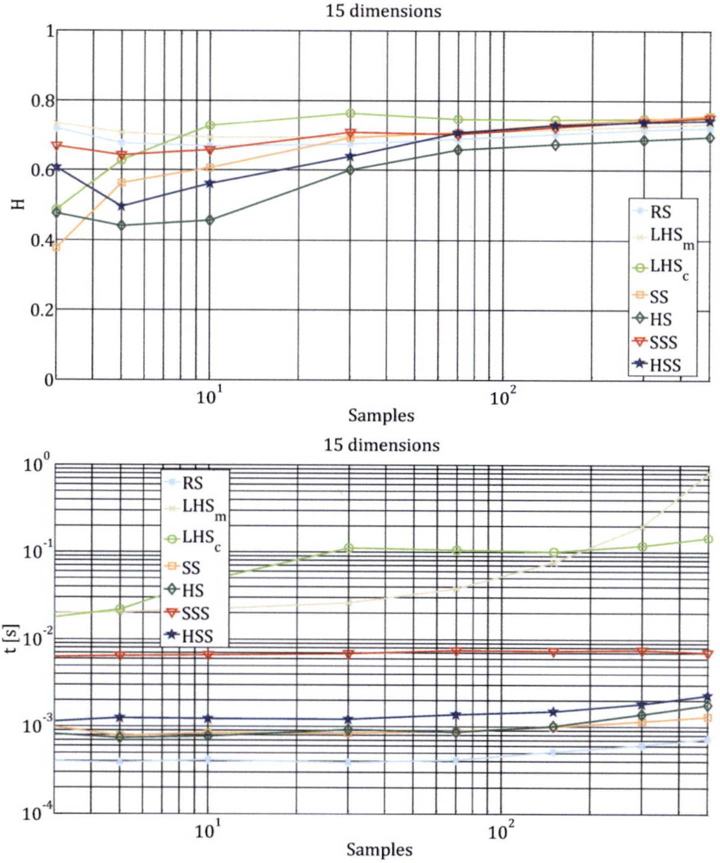

Figure 3.19: Comparison in 15 dimensions

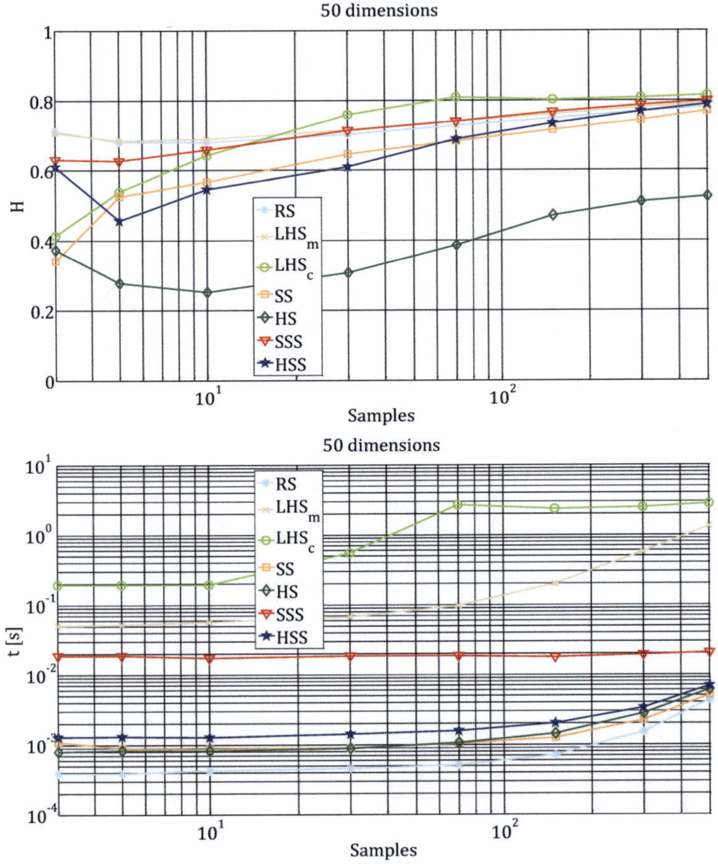

Figure 3.20: Comparison in 50 dimensions

Looking at the global homogeneity index, the choice of the design is crucial for few samples and high dimension: under these conditions the homogeneity of the design varies significantly. On the contrary, the quality of their distribution tends towards random distribution for an increasing number of samples. Regarding generation time, it increases over dimension and sample size, as expected. Moreover, LDS have "drops of quality" over dimensions: this, due to their deterministic formulation. For a given number of dimensions the sequence needs a minimum number of samples in order to be well distributed (Morokoff & Caflisch, 1994), otherwise strong correlations between the variables of two dimensions rise: examples are presented in Figure 3.21. In fact, the global homogeneity index of this HS is very low (see Figure 3.20).

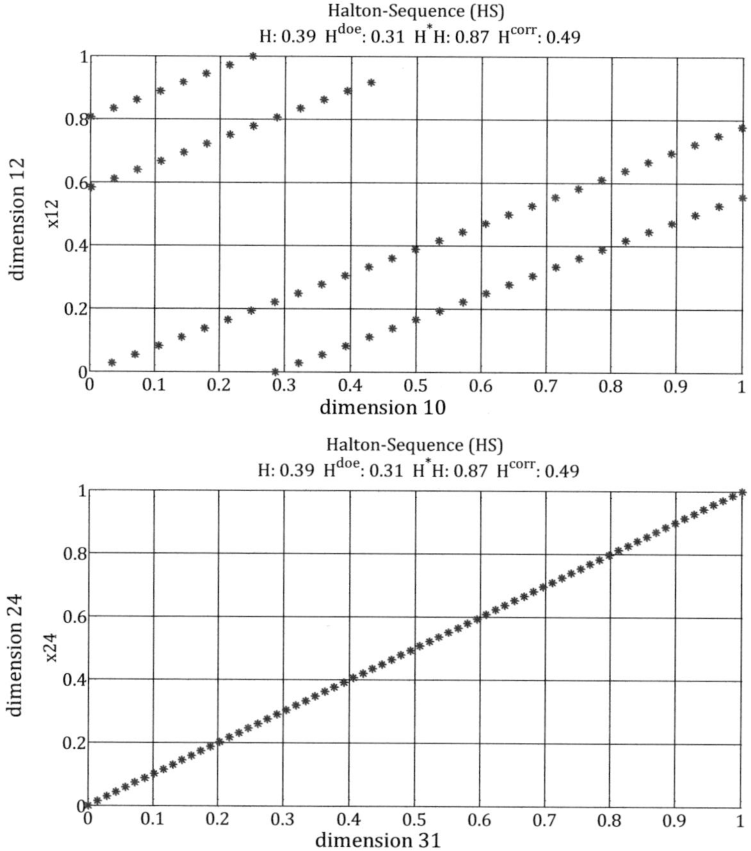

Figure 3.21: Halton Sequence in 50 dimensions with 70 samples

A graphical representation of the results allows to choose rapidly the most effective design: all DoE generated for the present thesis are chosen according to the introduced global homogeneity index. Generally, LHS outperform other designs: LHS_m are recommended for few samples ($n < 10$) and LHS_c for many samples ($n > 30$). The higher the dimension, the more LHS_m is closer to random sequence. However, LHS require more time to generate the design; but, as already mentioned, is often negligible if compared to the time necessary to evaluate the experiments set by the design. Concerning LDS, SS is recommendable for higher ($d > 7$), HS for lower dimensions ($d < 7$). However, both suffer from "drops" of quality. Permutation algorithms, like Matousek's scramble and RR2 scramble, improve the homogeneity of the distribution, especially for few samples and high dimensions. They are recommended above dimension 15. However, the improvement is guaranteed only on average: it could be necessary to repeat the generation in order to have a significant increase of homogeneity compared to the LDS without permutation algorithm.

3.5 Correlation Analysis

The generic term "correlation analysis" refers to a series of methods concerning numerical and graphical analyses, whose purpose is the identification of tendencies and influences between design variable and design solution parameter of a study.

In the present thesis, correlation analyses are used for:

- identifying if two parameters correlate positively or negatively;
- quantify the influence of one parameter on another;
- identify and represent graphically the relation between two parameters.

For the first application Pearson's and Spearman's correlation indices are used. To identify the strength of the influence Sobol's global sensitivity index is chosen. The graphical representation is performed using scatter plots with a trend-line calculated through a moving average algorithm.

3.5.1 Correlation Analysis

Correlation analyses quantify the type of the relation between design variables and design solutions, but not its strength. Person's and Spearman's correlation coefficients are well-known and widely established methods for correlation analyses: they quantify the linearity (Pearson) or monotonicity (Spearman) of the relation between design variables and design solutions. Moreover, if applied among the design solution variables, correlation analyses serve to reduce their number: if the correlation between two design solution variables is high, one can be disregarded.

In this section, Pearson's and Spearman's correlation index are briefly presented. A comparison between the two coefficients, as well as information about their history can be found in (Hauke & Kossowski, 2011)

3.5.1.1 Pearson's Correlation Coefficient

Pearson's correlation coefficient, introduced first by (Bravais, 1847) and standardized then by (Pearson, 1896), quantifies the linearity of the relation between two variables:

$$\rho_P = \frac{cov(\boldsymbol{x},\boldsymbol{y})}{\sigma_x\,\sigma_y} \tag{3.42}$$

where cov is the covariance, σ the standard deviation, $\boldsymbol{x}$ and $\boldsymbol{y}$ two variables.

To determine Pearson's correlation coefficient, the variables $\boldsymbol{x}$ and $\boldsymbol{y}$ have to be normally distributed and should at least belong to an interval scale, i.e. it should be allowed to measure the difference between two data points. For interval scales the calculation of mode, median and arithmetic mean is statistically allowed. In order to determine the statistical significance of Pearson's correlation analyses, the p-value test is used. If smaller than a given significance level (generally 5%) the null-hypothesis can be rejected, i.e. the correlation is statistically significant.

3.5.1.2 Spearman's Rank Correlation Coefficient

Spearman's correlation coefficient quantifies the monotony of the relation between two variables (Spearman, 1904):

$$\rho_S = \frac{cov(R(\boldsymbol{x}), R(\mathbf{y}))}{\sigma_{R(x)}\ \sigma_{R(y)}} \quad \textbf{3.43}$$

where cov is the covariance, σ the standard deviation, $R(\ldots)$ the rank of the values of one variable, $\boldsymbol{x}$ and $\boldsymbol{y}$ the two variables.

To determine Spearman's correlation coefficient, the variables $\boldsymbol{x}$ and $\boldsymbol{y}$ should at least belong to an ordinal scale: it should be allowed to rank the data, but not necessarily to measure the difference between two data points. For interval scales only the calculation of mode and median are statistically allowed. As the design variables and design solutions are ranked, Spearman's correlation coefficient is more robust against outliers.

In order to determine the statistical significance of Spearman's correlation analyses, the p-value test is used. If smaller than a given significance level (generally 5%) the null-hypothesis can be rejected: i.e. the calculated correlation coefficient is statistically significant.

3.5.2 Sensitivity Analysis

Sensitivity analyses quantify the strength of the relation between design variables and design solutions, but not its type. The choice and implementation of sensitivity analyses are crucial, as they influence the results: (Frey & Patil, 2002) presents a review of sensitivity analyses whereas (Ascough II, Green, Ma, & Ahjua, 2005) enlists key criteria for selecting and implementing them. Sensitivity analyses can be classified in different ways (Frey & Patil, 2002):

- local or global,
- mathematical, statistical or graphical.

For the present thesis, the focus lies on global sensitivity analyses (GSA) based on mathematical approaches. The advantage of GSA is to provide results that are valid for the whole design variable space. The most famous and sophisticated GSA are variance-based: e.g. the Analysis of Variance (ANOVA) (Archer, Saltelli, & Sobol, 1997) and Sobol's sensitivity analysis (Sobol, 1993). The result of a sensitivity analysis is summed up by a coefficient: it determines the influence of one design variables on a design solution, the so-called effect size. The effect size should be (Cohen, 1988):

- a non-dimensional number
- independent from the measurement unit of the data,
- independent from the number of samples,
- equal to zero, if the null hypothesis is wrong (if present).

Most sensitivity analyses provide more than one coefficient to measure the main-, side-, and total effect separately: the main effects determine the influence on a design solution related to only one design variable, side effects describe the influence on a design solution due to interactions between two or more design variables; total effects are the sum of main- and side effects.

3.5.2.1 Sobol's Global Sensitivity Analysis

In the literature, a very well-known analysis is Sobol's variance-based sensitivity analysis (Sobol, 1993). It is important to distinguish between Sobol's first-, second- and higher-order coefficients (see Table 3.22).

Table 3.22: Sobol's sensitivity coefficients

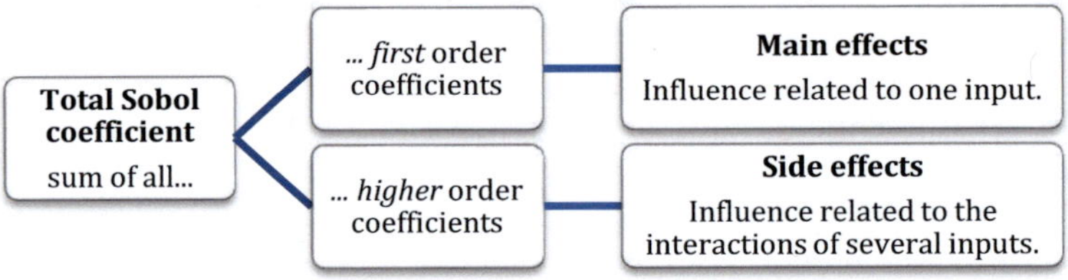

The variance σ^2 of the design solution $\boldsymbol{Y}$ due to the design variable $\boldsymbol{x}_j$ is:

$$\sigma_j^2 = Var\left(E\left(\boldsymbol{Y}|\boldsymbol{x}_j\right)\right) \qquad \forall\, j = 1, \dots, n_{DV} \qquad \mathbf{3.44}$$

The variance σ^2 of the design solution $\boldsymbol{Y}$ due to the combinations of two or more design variables is:

$$\sigma^2_{\underbrace{j,k,l,\dots}_{I\ indices}} = Var\left(E\left(\boldsymbol{Y}| \underbrace{\boldsymbol{x}_j, \boldsymbol{x}_k, \boldsymbol{x}_l, \dots}_{I\ indices} \right)\right) - \sum_{i=1}^{I-1} \sum_{jj=1}^{n_{DV}-i+1} \sum_{kk=jj+1}^{n_{DV}-i+2} \sum_{\dots}^{\dots} \sigma^2_{\underbrace{jj,kk,ll,\dots}_{i\ indices}} \qquad \mathbf{3.45}$$

where I defines the number of design variables $I \leq n_{DV}$ and the indices are: $\forall\, j = 1, \dots, n_{DV} - I + 1, \forall\, k = \mathrm{j} + 1, \dots, n_{DV} - I + 2$ and so on[3]. As $\boldsymbol{Y}$ is a function of all design variables $\boldsymbol{x}$, it is not possible to calculate analytically the variance σ^2: it has to be estimated (Gobbi, Mastinu, & Miano, 2005) (Han, 2011).

[3] In order to determine the variance $\sigma_{1,2,3}^2$ first the variance due to the variables $\boldsymbol{x}_1$, $\boldsymbol{x}_2$ and $\boldsymbol{x}_3$ is calculated and then the variances σ_1^2, σ_2^2, σ_3^2, $\sigma_{1,2}^2$, $\sigma_{1,3}^2$ and $\sigma_{2,3}^2$ are subtracted.

After having estimated all variances, it is possible to determine Sobol's first and higher order coefficients for each design variable. Sobol's first-order coefficients ($I = 1$) measure the main effects:

$$S_j^M = S_j^1 = \sigma_j^2/Var(\boldsymbol{Y}) \qquad \forall j = 1, \dots, n_{DV} \tag{3.46}$$

Sobol's higher order coefficients ($I \geq 2$) measure the side effects:

$$S^I_{\underbrace{j,k,l,\dots}_{I\ indices}} = \sigma^2_{\underbrace{j,k,l,\dots}_{I\ indices}} /Var(\boldsymbol{Y}) \tag{3.47}$$

Sobol's total coefficients quantifies the total influence (main and side effects) one design variable has on one design solution. It is calculated adding all Sobol's coefficients:

$$S_j^T = \sum_{I=1}^{n_{DV}} \sum_{k,l,\dots} S^I_{\underbrace{j,k,l,\dots}_{I\ indices}} \qquad \forall j = 1, \dots, n_{DV} \tag{3.48}$$

Sobol's sensitivity analysis gives a precise but not rapid evaluation of the design variables' sensitivity. In fact, in order to determine main and side effects it is necessary to have more samples[4] than those of the original DoE. If the mathematical description of the examined model is known, it could be used. On the contrary, if the equations of the model are unknown or the required time to evaluate them is impracticable, a meta model (e.g. artificial neural network) has to be generated in order to approximate the original system. In this case the precision and confidence of the sensitivity analysis depends on the quality of the meta model.

In the present thesis, for all sensitivity analysis Sobol's GSA is chosen, as it allows a global sensitivity evaluation, distinguishing between Sobol's

[4] For a DoE consisting of n samples and n_{DV} design variables the evaluations are $n \cdot (n_{DV} + 1)$ (Gobbi, Mastinu, & Miano, 2005).

main index and Sobol's total index. In general, the higher is Sobol's total index, the stronger is the influence of the design variables on the design solutions.

3.5.2.2 Graphical Sensitivity Analysis

Graphical sensitivity analyses are used if the samples number is too low compared to the complexity of the analysed system: in this case a numerical sensitivity analysis would not give reliable results.

Graphical analyses allow intuitive examinations using graphs, charts, or surfaces (Frey & Patil, 2002). Often they serve as screening method before further analyses, but they may also be implemented afterwards in order to verify the numerical results of statistical or mathematical analyses (e.g. (Stiber, Pantazidou, & Small, 1999)).

A typical graphical analysis is based on scatter-plots (Galvao, Pizarro, & Epiphanio, 2001): many programs (Optimus, S-Plus) generate them automatically. Scatter-plots allow to analyse the general trend between inputs and outputs: a trend-line helps to identify the tendency. In the present thesis, the trend-line is calculated as trough a moving average algorithm (Peckelsen, 2012).

4 Power Loss

4.1 Modelling Power Loss

In this section, first, the definition of power loss is introduced and its formulation for an outer drum test bench is presented. Then, the parameterization method for Magic Formula is described.

4.1.1 Definition of Power Loss

In this section the definition of power loss is derived from equations describing the mechanics and thermodynamics of a system; then, its formulation for an outer drum test bench is presented. For further details, especially concerning the validation procedure, refer to (Peckelsen, Gauterin, & Unrau, 2015).

4.1.1.1 General Formulation

Figure 4.1 shows the forces and torques acting on a driven tyre rolling at constant velocity. The entities s and v represent respectively the linear displacement and the velocity of wheel hub; ϑ and ω are the angular displacement and the angular velocity of the tyre. M_{mot} is the torque generated by a motor that keeps the tyre rolling. Tyre load is described by F_z, while the resulting longitudinal and lateral forces acting in the contact area parallel to the road are represented by F_x and F_y. M_r is the rolling resistance torque: it represents all losses caused by asymmetric pressure distribution, micro-slips in the tyre tread and hysteretic behaviour of the tyre compound. Force F_{aero} and torque M_{aero} are introduced to model the asymmetric distribution of aerodynamic forces.

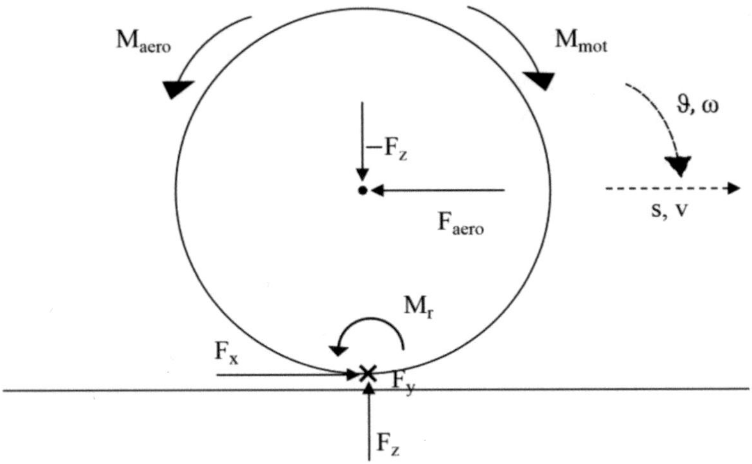

Figure 4.1: Driven tyre on a flat surface (Peckelsen, Gauterin, & Unrau, 2015)

Considering global forces and displacements, the principle of conservation of energy for a system is:

$$\Delta E = \Delta E_m + \Delta U = W - Q \qquad \textbf{4.1}$$

where ΔE is the variation of the total energy of the system, ΔE_m the variation of mechanical energy, ΔU the variation of internal energy, W the mechanical work (positive, if energy is introduced into the system), and Q the heat energy (positive, if heat is transferred to the ambient). Mechanical work is defined as:

$$W = M_{mot} \cdot \Delta\vartheta \qquad \textbf{4.2}$$

The sum of internal energy variation ΔU and energy loss Q due to heat exchange with the environment is equal to the total energy loss E_{loss} generated by the tyre. On the one hand, its generation can be attributed to forces and torques working for relative displacements or rotations (Q_r); on the other hand energy loss is caused by hysteresis of the tyre compound (Q_h):

$$E_{loss} = \Delta U + Q = Q_r + Q_h \qquad \textbf{4.3}$$

Energy loss Q_r is caused by slip, abrasive wear and aerodynamics. Q_r is generated by relative displacements (x_r, y_r) and rotations (camber angle γ_r, rotation angle ϑ_r, steering angle ψ_r) between tyre tread and road surface, and by relative displacements (s_r) or rotation (ϑ_r) between tyre and air:

$$\begin{aligned} Q_r = & \int_{x_r} F_x dx_r + \int_{y_r} F_y dy_r + \int_{\gamma_r} M_x d\gamma_r + \int_{\vartheta_r} M_y d\vartheta_r \\ & + \int_{\psi_r} M_z d\psi_r + \int_{s_r} F_{aero} ds_r + \int_{\vartheta_r} M_{aero} d\vartheta_r \end{aligned} \quad \mathbf{4.4}$$

Energy loss Q_h is caused by hysteresis of the compound of the tyre. Q_h is generated by non-elastic deformations (compression and expansion) of the tyre given by relative displacements (x_d, y_d, z_d) and rotations (γ_d, ϑ_d, ψ_d) between tyre tread and wheel hub. The amount of the non-elastic part of the deformation is defined by the loss factors c_{Fx}, c_{Fy}, c_{Fz}, c_{Mx}, c_{My}, c_{Mz}:

$$\begin{aligned} Q_h = & \int_{x_d} c_{F_x} F_x \, dx_d + \int_{y_d} c_{F_y} F_y \, dy_d + \int_{z_d} c_{F_z} F_z \, dz_d \\ & + \int_{\gamma_d} c_{M_x} M_x \, d\gamma_d + \int_{\vartheta_d} c_{M_y} M_y \, d\vartheta_d + \int_{\psi_d} c_{M_z} M_z \, d\psi_d \end{aligned} \quad \mathbf{4.5}$$

Power is defined as energy variation over time:

$$\dot{E}_m + \dot{U} = \dot{W} - \dot{Q} \quad \mathbf{4.6}$$

The power loss P_{loss} is then the sum of internal power $\dot{U}$ and heat flow $\dot{Q}$ (tyre abrasive wear is disregarded):

$$P_{loss} = \dot{U} + \dot{Q} = \dot{W} - \dot{E}_m \quad \mathbf{4.7}$$

4.1.1.2 Formulation for an Outer Drum Test Bench

Figure 4.2 shows acting forces and torques during a measurement on an outer drum test bench.

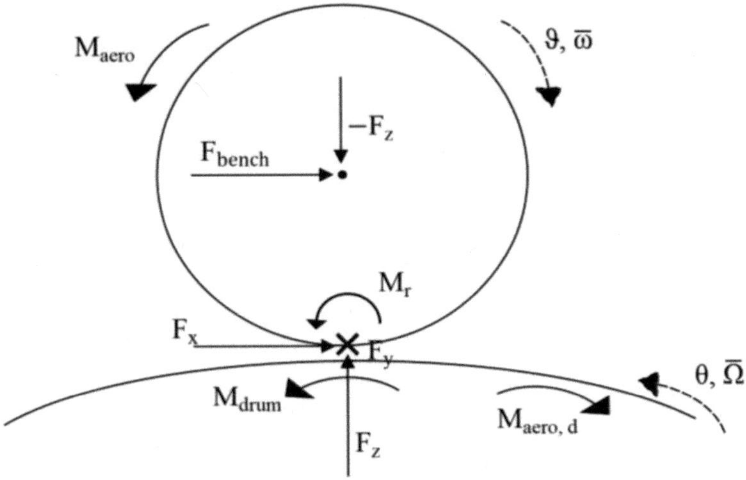

Figure 4.2: Rolling tyre on an outer drum test bench (Peckelsen, Gauterin, & Unrau, 2015)

The free rolling tyre (radius R) is driven at constant velocity by a motorized drum (radius R_d, drum torque M_{drum}). $\bar{\Omega}$ is the angular velocity of the drum and $\bar{\omega}$ the angular velocity of the tyre. The horizontal force of the drum test bench at the rim hub is F_{bench}.

At steady state condition, i.e. constant drum angular velocity $\bar{\Omega}$ ($\dot{E}_m = 0$), constant forces and torques at the wheel hub, no variations of camber γ and steering angle ψ and constant tyre temperature $\bar{T}$ ($\dot{U} = 0$), mechanical and internal power are zero. Under these conditions, all contributions of camber torque M_x and aligning torque M_z as well as aerodynamic force F_{aero} and non-elastic deformations in longitudinal and lateral direction can be disregarded (Peckelsen, Gauterin, & Unrau, 2015). Subsequently, power loss can be reduced to (drum aerodynamics are disregarded):

$$P_{loss} = \int_{x_r} F_x d\dot{x}_r + \int_{y_r} F_y d\dot{y}_r + \int_{\vartheta_r} M_y d\dot{\vartheta}_r \quad \textbf{4.8}$$

$$+\int_{z_d} c_{F_z} F_z \, d\dot{z}_d + \int_{\vartheta_d} c_{M_y} M_y \, d\dot{\vartheta}_d + \int_{\vartheta_r} M_{aero} d\dot{\vartheta}_r$$

If measurements are depurated by parasitic losses (i.e. frictional bearings and aerodynamic losses) as required by (ISO 28580, 2009), power loss can be expressed as:

$$P_{loss} = F_x(R\bar{\omega} - cos(\alpha)\bar{v}) - F_y \, sin(\alpha)\bar{v} + M_r\bar{\omega} \qquad \textbf{4.9}$$

where α is the slip angle (when positive F_y is by definition negative), $\bar{v} = R_d\bar{\Omega}$ the trajectory velocity and M_r includes non-elastic deformations due to spin torque F_z and vertical force M_y. If measurements are performed without braking or driving torques (i.e. zero longitudinal slip), equation 4.9 can be rewritten as:

$$P_{loss} = -F_y \, sin(\alpha)\bar{v} + M_r\bar{\omega} \qquad \textbf{4.10}$$

Equation 4.10 describes the power loss of a free rolling tyre at steady state conditions on an outer drum test bench. It is a function of velocity, slip angle and (indirectly through F_y and M_r) of camber angle and tyre load.

4.1.2 Parameterization Method for Magic Formula

In order to evaluate the power loss of a tyre modelled by the MF-model 5.2, it is necessary to parameterize both equations describing the lateral force F_y and the rolling resistance torque M_r. Subsequently, two different measurements are needed: first, the MF-model has to be fitted to TIME2 measurements (Tischleder, Köhne, Leister, & Bode, 2004), then, to rolling resistance measurements (ISO 28580, 2009). For both parameterizations the Tyre Fitter can be implemented (see section 3.3.2).

Figure 4.3: Used FlatTrac for TIME2 measurements

The TIME2 measurements are performed on a FlatTrac (see Figure 4.3) according to the standard TIME2 routine (Tischleder, Köhne, Leister, & Bode, 2004).

The rolling resistance measurements are performed on an outer drum test bench (see Figure 4.4) according to the standard ISO routine (ISO 28580, 2009), but for different tyre loads and velocities (see Table 4.5).

Figure 4.4: Used outer drum test bench for rolling resistance measurements

For all test sessions the same tyre is used. No thermodynamic correction is applied for the different velocities. It is assumed that the contribution of internal power $\dot{U}$ is negligible compared to the power loss generated by heat flow $\dot{Q}$ (see equation 4.7). Rolling resistance measurements are then transformed into power loss multiplying drum torque M_{drum} with drum angular velocity $\bar{\Omega}$. Validation measurements are presented in (Peckelsen, Gauterin, & Unrau, 2015).

Table 4.5: Measurements procedure for rolling resistance fitting

Test session 1	
Velocity	$80\ km/h$
Slip angle	0°
Camber angle	0°
Tyre load	$4000\ N$
Test session 2	
Velocity	$80\ km/h$
Slip angle	0°
Camber angle	0°
Tyre load	1000 to 6000 N (step 1000 N)
Test session 3	
Velocity	40 to 190km/h (step 30 km/h)
Slip angle	0°
Camber angle	0°
Tyre load	$4000\ N$

4.2 Objective Manoeuvre Criteria

In this section, first, the manoeuvre used to evaluate the power loss of tyres is presented; then, the objective manoeuvre criteria (OMC) necessary to describe the power loss characteristics are defined.

4.2.1 Choice of Manoeuvres

The manoeuvre is defined to reproduce best the customer average driving behaviour and its related power loss. According to equation 4.9, power loss can be influenced by the customer operating conditions: driving velocity, steering angle and longitudinal acceleration.

To define a virtual manoeuvre for describing power loss, data of over 50 BMW customers is analysed. The available data is based on measurements concerning different vehicles, drivers and European countries: it includes information about the time customers drive at a given driving velocity using a given steering wheel angle.

The histograms in Figure 4.6 and Figure 4.7 show the statistics concerning the percentage driving time at different driving velocities and steering wheel angles (Peckelsen, D'Avanzo, Bode, Brenker, & Gauterin, 2013).

Customers spend more than 50 % of driving time at velocities up to 100 km/h. The average driving velocity is approximately $80 km/h$. Moreover, customers do not use steering wheel angles bigger than 45 ° for almost 90 % of driving time.

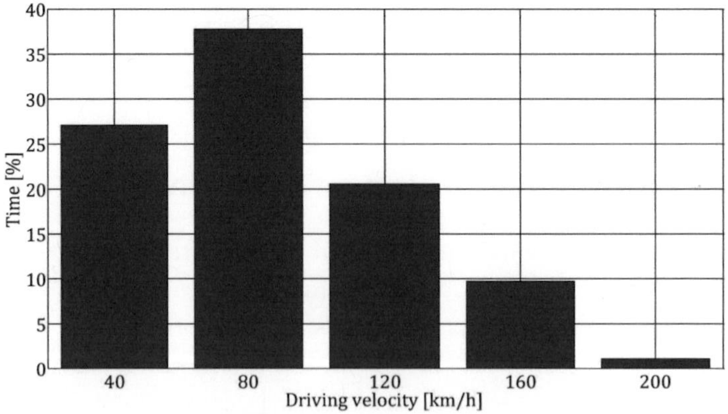

Figure 4.6: Customer statistics for driving velocity

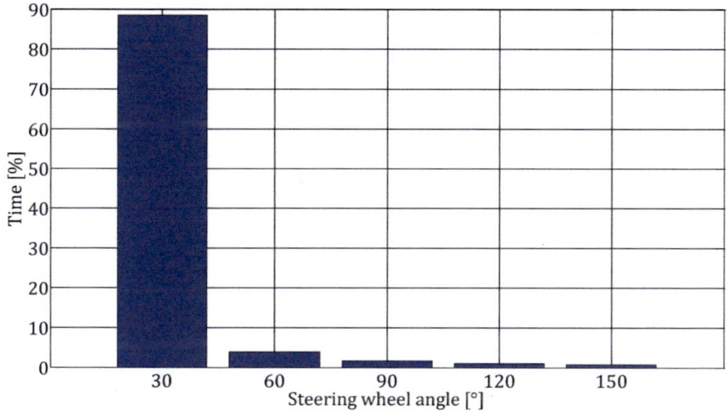

Figure 4.7: Customer statistics for steering wheel angle

Starting from this customer statistic a statistic for lateral acceleration is derived virtually using ramp steer manoeuvres (see section 0). In Table 4.8 the time percentages spent at each operating condition is shown: it is interesting to notice, that low accelerations (below 0.5 m/s) are reached at low velocities, medium accelerations at velocities typical of country roads (80 km/h); whereas high lateral acceleration (above 4.0 m/s) are reached again at lower velocities (below $40 km/h$), probably on roundabouts or sharp turns.

Table 4.8: Customer statistics for driving velocity and lateral acceleration

Driving velocity [km/h]	0	0.5	1.0	1.5	2	2.5	3.0	3.5	4.0	4.5	5.0	5.5
200												
160						1.13						
120				4.17	3.70							
80			13.48	3.97	0.50	0.36	0.18	0.17				
40	7.85	28.70	2.38	2.73	1.00	1.11	0.62	0.56	0.32	0.16	0.15	
0	15.42	3.71	1.16	0.86	0.75	0.68	0.53	0.45	0.10	0.18	0.13	0.11

Lateral acceleration [m/s²]

Table 4.9: Ramp steer, manoeuvre parameters

Name	Symbol	Range
Driving velocity	v	20 km/h, 60 km/h, 100 km/h, 140 km/h, 180 km/h,
Steering wheel angle	δ_H	0° ... x quasi-statically increasing till maximal lateral acceleration.

Knowing the statistic for driving velocity and lateral acceleration it is possible to define a set of virtual manoeuvres for representing the customer average power loss under real operating conditions. To this purpose a series of ramp steers (see section 0) is chosen. The manoeuvre parameters are enlisted in Table 4.9: the ramp steer is performed at five different velocities; the maximum steering wheel angle is adjusted so, that the maximum lateral accelerations for each velocity are reached. As vehicle and tyre model are symmetric, it is not necessary to perform the ramp steer steering both clockwise and counter-clockwise. This manoeuvre allows to evaluate virtually forces and torques at the tyre contact patch for a significant number of customer operating conditions.

4.2.2 Choice of Objective Manoeuvre Criteria

The chosen OMC is the "customer average power loss": it is the weighted sum of the power loss generated by all four tyres of a vehicle under real customer operating conditions. The power loss of a four tyre i under the operating condition j is (see equation 4.9):

$$P_{loss,ij} = F_{x,i}(R_i\overline{\omega}_i - cos(\alpha_i)\overline{v}_i) - F_{y,i}\,sin(\alpha_i)\overline{v}_i + M_{r,i}\overline{\omega}_i \qquad \textbf{4.11}$$

where i indicates the four tyres: front left, front right, rear left and rear right. The average power loss of all tyres for the operating condition j is:

$$P_{loss,j} = \sum\nolimits_i \frac{P_{loss,ij}}{4} \qquad \textbf{4.12}$$

The customer average power loss per tyre $\bar{P}_{loss}$ is then:

$$\bar{P}_{loss} = \sum_{j} w_{loss,j} \cdot P_{loss,j} \tag{4.13}$$

where the weight $w_{loss,j}$ represents the percentage time the customer spends at the operating condition j.

4.3 Functional Tyre Characteristics

In this section, first, the functional tyre characteristics (FTCs) describing the relevant characteristics of the power loss of tyres are defined; then, their influence on the OMC is analysed and the relevant FTCs are chosen.

4.3.1 Definition of Functional Tyre Characteristics

The FTCs should allow a unique and exhaustive description of the power loss of tyres. According to equation 4.9, power loss is a function of velocity $\bar{v}$, slip angle α, camber angle γ (indirectly through F_y and M_r), tyre load F_z (indirectly through F_x, F_y and M_r) and longitudinal slip κ:

$$P_{loss} = F_x(R\bar{\omega} - cos(\alpha)\bar{v}) - F_y\, sin(\alpha)\bar{v} + M_r\bar{\omega} \tag{4.14}$$

It is characterized by the expressions of longitudinal force, lateral force and rolling resistance torque. FTCs describing longitudinal and lateral force are presented in (Niedermeier, Peckelsen, & Gauterin, 2013). Subsequently, only FTCs describing rolling resistance torque M_r have to be added.

Given a tyre temperature and an inflation pressure, rolling resistance torque M_r is mainly a function of velocity, tyre load and longitudinal slip, as well as. Table 4.10 shows the FTCs necessary to describe M_{r0}, measured at reference conditions and its variation over velocity, tyre load, and longitudinal slip (Peckelsen, Gauterin, & Unrau, 2015).

Table 4.10: The four FTCs for power loss (defined at 4000 N tyre load and 80 km/h trajectory velocity)

Name	Formulation	Symbol	Unit
Rolling resistance torque	$M_r\|_{ref}$	$M_{r,0}$	$[Nm]$
Variation due to velocity	$\left.\frac{\Delta M_r/\|M_{r,0}\|}{\Delta v}\right\|_{ref} \cdot 100$	$M'_{r,0,v}$	$[\% \, s/m]$
Variation due to tyre load	$\left.\frac{\Delta M_r/\|M_{r,0}\|}{\Delta F_z}\right\|_{ref} \cdot 100$	$M'_{r,0,F_z}$	$[\%/N]$
Variation due to longitudinal slip	$\left.\frac{\Delta M_r/\|M_{r,0}\|}{\Delta \kappa}\right\|_{ref} \cdot 100$	$M'_{r,0,\kappa}$	$[\%/-]$

Within customer operating conditions, it is allowed to approximate real variation around $M_{r,0}$ with linear gradients. These FTCs can be added to the existing FTCs of the Tyre Shaper (see section 3.3.1), allowing the generation of tyres that have predefined power loss characteristics.

4.3.2 Choice of Functional Tyre Characteristics

The choice of the relevant FTCs is done according to a global sensitivity analysis (GSA) based on a design of experiments (DoE) consisting of 100 virtual tyres. The chosen manoeuvre is performed using the MF model 5.2 for the tyres (see section 3.2.2.1) and a double-track model for the vehicle.

The experiments of the DoE are virtual tyres generated through the Tyre Shaper (see section 3.3.1). The FTC variations are set to cover the ranges measured on real tyres. The design variables of the DoE are the four FTCs for power loss; the design solution is the average customer power loss. The number of virtual tyres is raised until the indices of the correlation and sensitivity analysis converged to a stable value: 100 virtual tyres represent the best trade-off between time and precision.

Table 4.11: GSA between the FTCs and OMC of power loss

	$\bar{P}_{loss}$ ($R^2 = 0.99$)
	T
$M_{r,0}$	**0.59**
$M'_{r,0,v}$	0.12
$M'_{r,0,F_z}$	0.18
$M'_{r,0,\kappa}$	0.11

Above 100 no significant improvements concerning the convergence of the indices is observed. The method for generating the DoE is chosen according to the homogeneity index (see section 3.4.2.4): the most homogeneous design for four design variables and 100 experiments is generated by a latin hypercube. Sobol's GSA is chosen, as it allows a global sensitivity evaluation (see section 3.5.2.1): the higher is Sobol's total index "T", the stronger is the influence of the FTCs on the OMC. Moreover, the R^2-value of the artificial neuronal networks (ANN) needed to determine Sobol's indices is 0.99: the results are statistically significant.

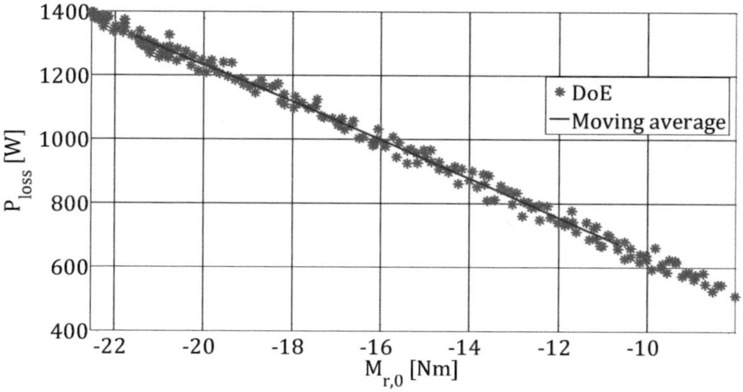

Figure 4.12: Graphical analysis for power loss

Table 4.13: Relevant FTC for power loss

Name	Formulation	Symbol	Unit
Rolling resistance torque at reference conditions	$M_r\|_{ref}$	$M_{r,0}$	$[Nm]$

The results of the analysis are shown in Table 4.11. Rolling resistance torque $M_{r,0}$ is the dominant FTC. Moreover, it should be noticed, that the influence of the variation due to velocity $M'_{r,0,v}$ is very low, although no thermodynamic correction is made on temperature.

The linear relation between power loss and rolling resistance torque $M_{r,0}$ is shown graphically in Figure 4.12: the low variance around the moving average proves that the other FTCs are of minor importance compared to rolling resistance torque $M_{r,0}$.

Rolling resistance torque $M_{r,0}$ is the dominant FTC and is chosen to characterize the power loss of tyres (see Table 4.13).

4.4 Summary

In section 1 the requirement power loss is analysed. First, a definition of power loss is derived according to equations describing the mechanics and thermodynamics of a system (see equation 4.7). A formulation for the evaluation of power loss on an outer drum test bench is derived (see equation 4.9).

Then, according to the definition of power loss one objective manoeuvre criteria is defined: the customer average power loss P_{loss} (see equation 4.13). Its evaluation is based on a series of ramp steer manoeuvres and on a customer statistics describing the time customers spend at different driving velocities and lateral accelerations.

Finally, four functional tyre characteristics (FTCs) describing the power loss of tyres under real vehicle-driving conditions are defined. Then, their influence on the objective manoeuvre criteria (OMC) is analysed and one relevant FTC for characterising the power loss of tyres is chosen: the rolling resistance torque $M_{r,0}$ (see Table 4.13).

5 Lateral Dynamics

5.1 Objective Manoeuvre Criteria

In this section, first, the manoeuvres used to evaluate the lateral dynamics of vehicles are presented; then, the objective manoeuvre criteria (OMC) necessary to describe the lateral dynamics are chosen according to an objectivation study at the dynamic driving simulator.

5.1.1 Choice of Manoeuvres

The manoeuvres for lateral dynamics are defined to evaluate best the lateral dynamics of vehicles from steady-state conditions to transient limit driving. The choice reflects the state of the art procedures in the tyre development industry and takes into account correlation analyses between the objective manoeuvre criteria (OMC) and the subjective manoeuvre indices (SMI).

Four manoeuvres are chosen: the "ramp steer" to characterize the steady-state behaviour up to maximal lateral acceleration, the "sweep" and the "linearity" to evaluate the vehicle reaction to a sinusoidal and to a step steering input at low-to-middle lateral acceleration, and the "lane change" to characterize the vehicle reaction and stability at middle-to-high lateral acceleration.

5.1.1.1 Ramp Steer

The ramp steer (see section 0) is used to characterize the steady-state behaviour of vehicles. It is preferred to the quasi steady-state cornering as it is an open-loop manoeuvre.

Table 5.1: Ramp steer, manoeuvre parameters

Name	Symbol	Value / Range
Driving velocity	v	80 km/h
Steering wheel angle	δ_H	0° ... x quasi-statically increasing until maximal lateral acceleration is reached.

In the present thesis, for the evaluation of lateral dynamics, the ramp steer is performed at constant velocity v and with a quasi-statically increasing steering wheel angle δ_H (see Table 5.1). The steering wheel angle is increased until maximal lateral acceleration is reached. The ramp steer is mainly used to evaluate the characteristics of sideslip angle β, lateral acceleration a_y and roll angle φ as a function of steering wheel angle δ_H.

5.1.1.2 Continuous sine sweep

The continuous sine sweep (see section 2.2.1.2) is used to characterize the vehicle reaction to a sinusoidal steering input at low-to-middle lateral accelerations. On test tracks, the continuous sine sweep is performed as a closed-loop manoeuvre: the driver has to adjust the steering wheel angle in order to avoid the vehicle to drift sideward. In simulation, if vehicle and tyre models are symmetric and if tyres are excited only in their linear range (low-to-middle lateral accelerations), the vehicle drift is negligible and the continuous sine sweep can be performed as an open-loop manoeuvre (our case).

In the present thesis, for the evaluation of lateral dynamics, the continuous sine sweep is performed at constant velocity v; the input is a sinusoidal steering wheel angle with a constant frequency f_{δ_H} and increasing amplitude (see Table 5.2). The steering wheel angle amplitude is adjusted so, to reach lateral accelerations a_y of circa 2 m/s^2 (see Figure 5.3).

Table 5.2: Continuous sine sweep, manoeuvre parameters

Name	Symbol	Value / Range
Driving velocity	v	$100\ km/h$
Steering wheel angle frequency	f_{δ_H}	$0.5\ Hz$
Lateral acceleration	a_y	0 to $2\ m/s^2$

5.1.1.3 Linearity

The so-called linearity is composed of two step steering inputs (see section 2.2.1.3) and it is used to characterize the vehicle reaction at low-to-middle lateral accelerations. Similarly to the continuous sine sweep, the linearity is a closed-loop manoeuvre on test tracks, but can be performed as an open-loop manoeuvre in simulation, if vehicle and tyre models are symmetric and if tyres are excited only in their linear range (our case).

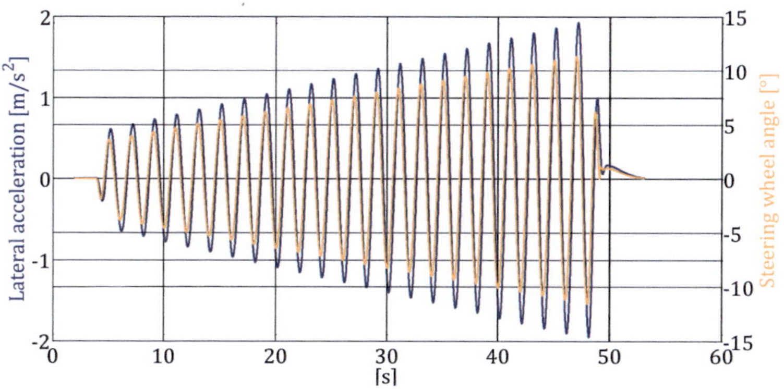

Figure 5.3: Continuous sine sweep, steering wheel angle and lateral acceleration (simulated with a double-track model and the MF model 5.2)

Table 5.4: Linearity, manoeuvre parameters

Name	Symbol	Value / Range
Driving velocity	v	$160\ km/h$
Maximal lateral acceleration	a_y	$3\ m/s^2$

In the present thesis, for the evaluation of lateral dynamics, the linearity is performed at constant velocity v. The inputs are two step steering input. The steering wheel angle amplitude is adjusted in a pre-manoeuvre so, that a given lateral acceleration a_y of circa $3\ m/s^2$ is reached (see Table 5.4). The given steering wheel angle amplitude is reached after $0.5\ s$, and then held for $2.2\ s$ (see Figure 5.5): this "holding phase" allows the vehicle to reach steady-state conditions before the start of the next step steering input.

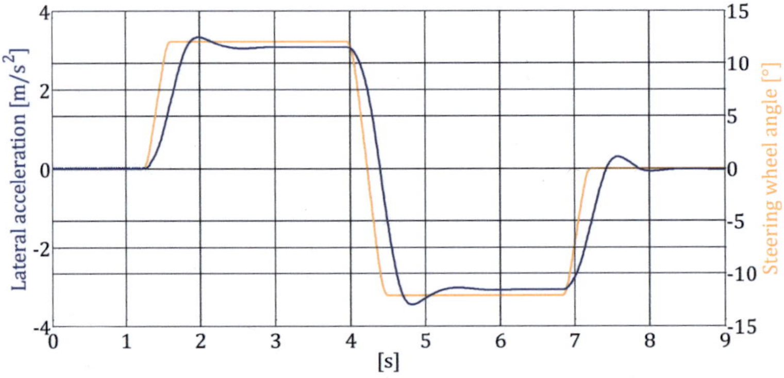

Figure 5.5: Linearity, steering wheel angle and lateral acceleration (simulated with a double-track model and the MF model 5.2)

5.1.1.4 Lane Change

The so-called lane change is composed of two step steering inputs (see section 2.2.1.3) and it is used to characterize the vehicle reaction at middle-to-high lateral accelerations. It is used to evaluate the vehicle stability as well as the time delay between vehicle reaction and steering input. Similarly to the linearity, the lane change can also be performed as an open-loop manoeuvre: however, due to the middle-to-high lateral accelerations, tyres are excited in their non-linear range causing the vehicle to drift sideward.

Table 5.6: Lane change, manoeuvre parameters

Name	Symbol	Value / Range
Driving velocity	v	160 km/h
Maximal lateral acceleration	a_y	6 m/s^2

Similarly to the linearity, the lane change is performed at constant velocity v; the inputs are two step steering. However, the steering wheel input is faster and has a bigger amplitude; subsequently, the vehicle reaction is higher. The steering wheel angle amplitude is adjusted for each tyre-vehicle combination reaching a lateral acceleration a_y of circa 6 m/s^2 (see Table 5.6). The given steering wheel angle amplitude is reached after 0.3 s, and then held for 1.2 s (see Figure 5.7): this "holding phase" allows the vehicle to reach steady-state conditions. The steering input can be repeated alternatively steering clockwise and counter-clockwise.

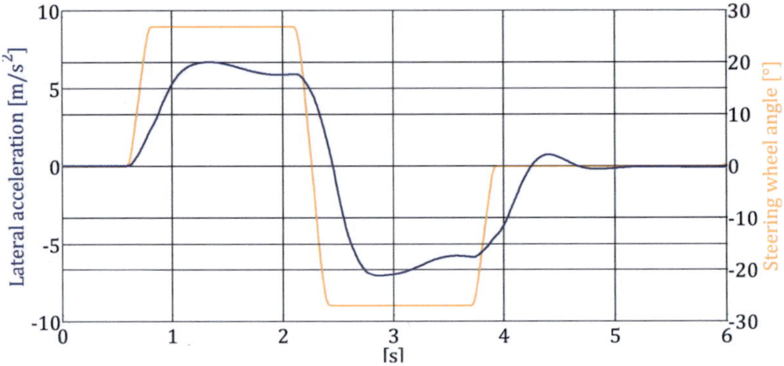

Figure 5.7: Lane change, steering wheel angle and lateral acceleration (simulated with a double-track model and the MF model 5.2)

5.1.2 Vehicle Dynamics Formulae

In this section, a new set of equations for vehicle dynamics are derived, hereinafter called "Vehicle Dynamics Formulae" (VDF). The proposed set of equations allows the analytical modelling of a transient vehicle response to a steering input without using vehicle or tyre models: the movements of the vehicle, modelled as a single rigid body, are described by the vehicle's degrees of freedom (DoF). Although the VDF are analytical equations with no physical meaning, the resulting movements are coherent to the physics of rigid body dynamics (Diana & Cheli, 2005).

The VDF have two main advantages: first, they allow to change the reaction of single DoF, keeping all other vehicle responses constant. For example, the roll angle reaction can be varied without changing the lateral acceleration response and the related time delay. Second, knowing the necessary analytical equations for modelling the vehicle reaction (e.g. the lateral dynamics in a lane change) allows to derive a unique set of possible OMC. The „possible OMC" represent the initial set of OMC: they are defined before performing the objectivation study. Sensitivity analyses will then identify the most sensitive OMC, which are a subset of the initial set of "possible OMC".

5.1.2.1 Formulae

The VDF presented in this section focus on the modelling of the vehicle response for a linearity or lane change manoeuvre as described in section 5.1.1. Both manoeuvres are characterized by:

- constant driving velocity,
- two step steering wheel angle inputs,
- a holding phase to allow the vehicle to reach steady-state conditions,
- absence of environmental excitations (e.g. wind, road surface irregularities).

Under these conditions, it is possible to:

- model the vehicle as a single rigid body,
- neglect longitudinal dynamics (i.e. longitudinal acceleration and pitch angle as the longitudinal movement of the vehicle is given by the constant trajectory velocity),
- neglect vertical dynamics (i.e. bouncing accelerations).

Subsequently, the six DoF of a vehicle are reduced to three: the independent variables sideslip angle β, lateral acceleration a_y and roll angle φ are chosen as most significant for the description of the vehicle reaction[5]. All other vehicle displacements, velocities and accelerations can be derived according to the physics of rigid body dynamics (Diana & Cheli, 2005).

[5] Yaw velocity $\dot{\psi}$ may be preferred to sideslip angle β. However, this implies a numerical optimisation, as $\dot{\psi} = f(\beta, \dot{\beta})$.

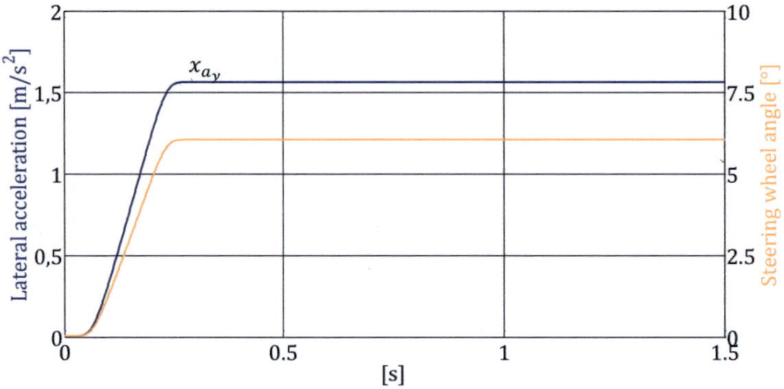

Figure 5.8: VDF, gain function

The VDF are based on the idea, that the vehicle response can be modelled through a "gain function", a "delay function" and an "overshoot function".

The first function characterizes the strength of the response at steady-state conditions. The second function defines at each time step the time delay between vehicle response and steering input. The third function describes the overshoots (oscillations) that occur before reaching steady-state conditions.

The "gain function" is (see exemplarily Figure 5.8):

$$x_i(t) = g_i \cdot \frac{\delta_H(t)}{\delta_{H,max}} \qquad \forall i \in [\beta, a_y, \varphi] \qquad \textbf{5.1}$$

where g_i is the gain factor, δ_H the steering wheel angle input, and $\delta_{H,max}$ the steering wheel angle amplitude reached at steady-state conditions.

The "delay function" models the time delay between vehicle reaction and steering input:

$$y_i(\hat{x}_i(t)) = \begin{cases} a_{i,1}\, \hat{x}_{i,1}^{b_{i,1}} + c_{i,1} \\ a_{i,2}\hat{x}_{i,2} + c_{i,2} \\ a_{i,3}\hat{x}_{i,3}^{b_{i,2}} + c_{i,3} \end{cases} \quad \begin{array}{c} \hat{x}_{i,1} \in [0, k_{i,1}] \\ \hat{x}_{i,2} \in [k_{i,1}, k_{i,2}] \\ \hat{x}_{i,3} \in [k_{i,2}, 1] \\ \forall i \in [\beta, a_y, \varphi] \end{array} \qquad \textbf{5.2}$$

where $a_{i,j}$, $b_{i,j}$ and $c_{i,j}$ are eight factors describing slope, shape and offset of the delay function, $k_{i,j}$ are factors defining the transition points, and $\hat{x}_{i,j}$ are the normalized gain functions:

$$\hat{x}_i(t) = \frac{x_i(t)}{\max(x_i)} \qquad \forall i \in [\beta, a_y, \varphi] \qquad \textbf{5.3}$$

The boundary conditions guarantee a continuously differentiable delay function:

$$y_{i,2}(k_{i,1}) = y_{i,1}(k_{i,1})$$
$$y_{i,3}(k_{i,2}) = y_{i,2}(k_{i,2})$$

$$\frac{dy_{i,2}(k_{i,1})}{d\hat{x}} = \frac{dy_{i,1}(k_{i,1})}{d\hat{x}} \qquad \forall i \in [\beta, a_y, \varphi] \qquad \textbf{5.4}$$
$$\frac{dy_{i,3}(k_{i,2})}{d\hat{x}} = \frac{dy_{i,2}(k_{i,2})}{d\hat{x}}$$

In Figure 5.9 the delay function of the lateral acceleration is exemplarily shown: 0.2 s time delay at 60 % means, that 60 % of the lateral acceleration is reached 0.2 s later than 60 % of the steering wheel angle input.

The "overshoot function" models the first and second oscillation of the vehicle reaction (see Figure 5.10). The overshoot function is modelled as:

$$\tilde{x}_i = p_{i,1} + p_{i,2} - (p_{i,1}(0) + p_{i,2}(0)) \qquad \forall i \in [\beta, a_y, \varphi] \qquad \textbf{5.5}$$

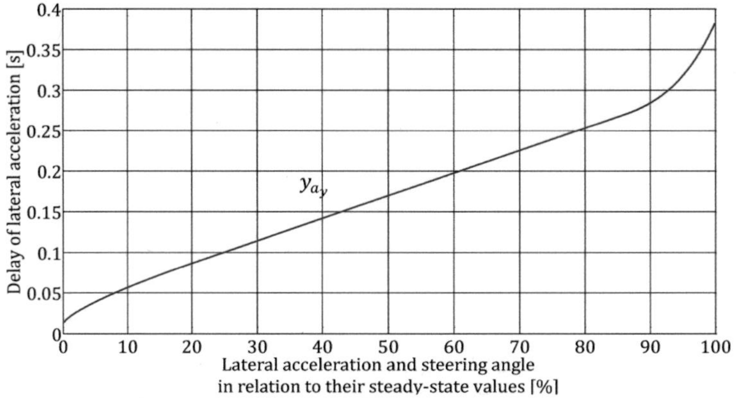

Figure 5.9: VDF, delay function

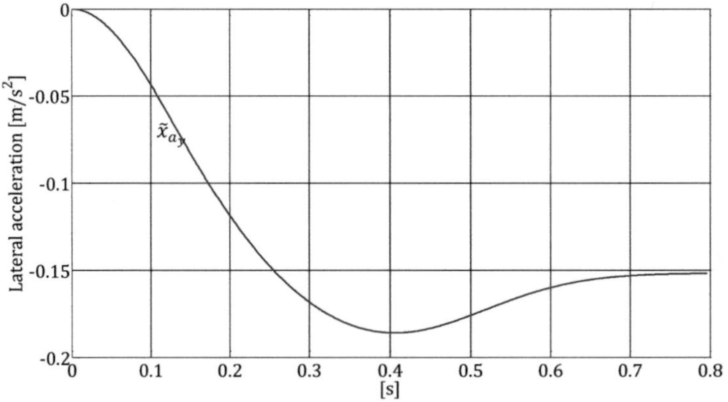

Figure 5.10: VDF, overshoot function

where the two peaks are approximated by a Gauss similar function, which is continuous and intently differentiable:

$$p_{i,1}(u_1) = \frac{\frac{d_{i,1}}{100} \cdot \max|x_i|}{\sqrt{2w_{i,1}\pi}} \cdot e^{\frac{-u_1^2}{2w_{i,1}}} \qquad \forall i \in [\beta, a_y, \varphi] \qquad \textbf{5.6}$$

$$p_{i,2}(u_2) = \frac{\frac{d_{i,2}}{100} \cdot \frac{d_{i,1}}{100} \cdot \max|x_i|}{\sqrt{2w_{i,1}\pi}} \cdot e^{\frac{-u_2^2}{2w_{i,1}}} \qquad \forall i \in [\beta, a_y, \varphi] \qquad \mathbf{5.7}$$

where $d_{i,j}$ are the percentage amplitudes of the peaks in relation to the amplitude of the vehicle reaction and $w_{i,j}$ the widths of the peaks (actually the variance of the Gauss function). The widths $w_{i,j}$ are set to $w_{i,1} = 0.3$ and $w_{i,2} = 0.2$.
The variables u_i are normalized time functions:

$$u_1 = 2 \cdot \frac{\tilde{t} - \tilde{t}(0)}{\max\left(\tilde{t} - \tilde{t}(0)\right)} \qquad \mathbf{5.8}$$

$$u_2 = 3 \cdot \frac{\tilde{t} - \tilde{t}(0)}{\max\left(\tilde{t} - \tilde{t}(0)\right)} - 1.5 \qquad \mathbf{5.9}$$

where $\tilde{t}$ is the time vector concerning the overshoot period. The normalization is performed so, to obtain $u_1 \in [0, 2]$ and $u_2 \in [-1.5, 1.5]$. The steering wheel angle signal is amplified using the gain function x_i. Then, it is shifted in time using the delay function y_i and the overshoot $\tilde{x}_i$ and its oscillation is added. These three stages are exemplarily shown in Figure 5.11.

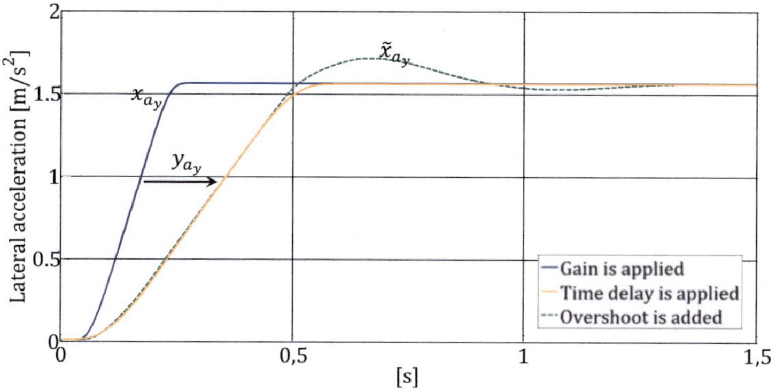

Figure 5.11: VDF, the three different stages

5.1.2.2 Case Study

In the following case study, the reliability of the VDF in reproducing a virtual reference vehicle dynamic is evaluated. The reference dynamic is performed using the MF model 5.2 for the tyres (see section 3.2.2.1) and a double-track model for the vehicle. First, a graphical comparison between VDF and reference vehicle simulation is shown; then, both models are evaluated by test drivers at the dynamic driving simulator.

The case study concerns the reproduction of a lane change performed by three different vehicles: a compact vehicle, a middle-class sport utility vehicle (SUV) and a high-class limousine. The parameters of the VDF are fitted, to best reproduce the reference simulation; an optimisation algorithm based on Matlab's "fmincon" is used.

The results concerning the modelling of the delay functions of the high-class limousine are shown in Figure 5.12: initial time delay and shape of the delay function are reproduced perfectly. The results concerning the characteristics over time of roll angle, sideslip angle and lateral acceleration are shown in Figure 5.13: again, shape, peak value and overshoots are approximated perfectly. Similar results are obtained for the compact vehicle and the middle-class SUV.

The lane changes, as performed by the three vehicles, are evaluated at the dynamic driving simulator (see section 5.1.3.1) by test drivers. The differences between VDF and reference simulation are negligible: the lateral dynamics of both models are evaluated equally good. Concluding, the presented VDF (see section 5.1.2.1) allow an accurate reproduction of the vehicle reaction to a step steering input as simulated by a double-track model and the MF model 5.2: not only the virtual results are comparable, but also the experienced dynamics are evaluated equally good.

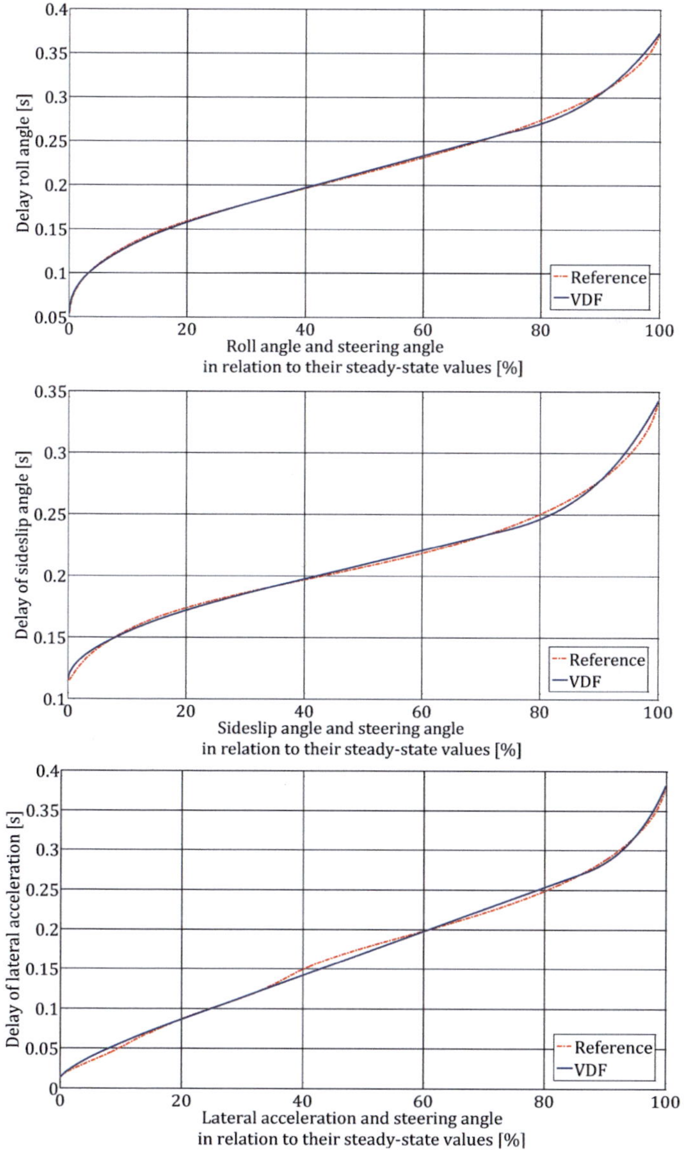

Figure 5.12: Case study, delay functions

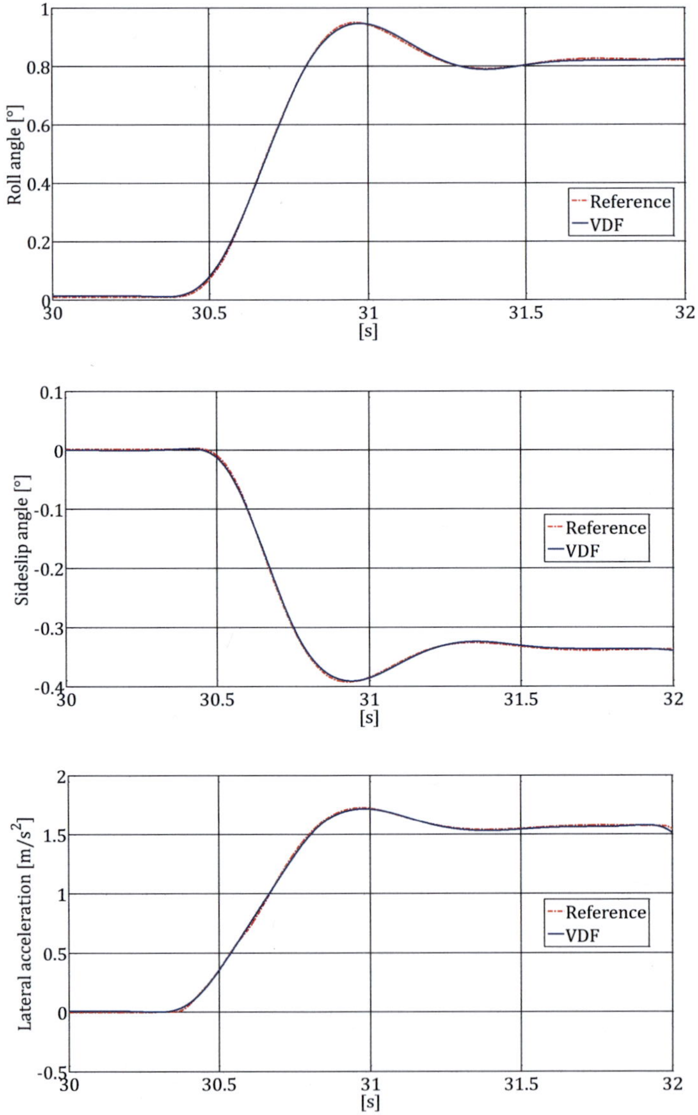

Figure 5.13: Case study, vehicle response

5.1.3 Objective Manoeuvre Criteria

In this section, first, the set of possible objective manoeuvre criteria (OMC) is motivated. Then, the DoE for the objectivation study and the evaluation procedure at the dynamic driving simulator are described. Finally, the results of the correlation analyses between the OMC and the subjective manoeuvre indices (SMI) are presented.

5.1.3.1 Definition of Objective Manoeuvre Criteria

Four manoeuvres are introduced for evaluating the influence tyres have on the lateral dynamics of vehicles: the ramp steer, the continuous sine sweep, the linearity and the lane change. In this section, for each manoeuvre the set of possible OMC is presented.

Table 5.14: Possible OMC for ramp steer

Characteristics	Formulation	Description
Steering wheel angle	$\delta_H\|_{4\ m/s^2}$	Steering wheel angle at a lateral acceleration of 4 m/s^2.
	$\left.\frac{\partial \delta_H}{\partial a_y}\right\|_{4\ m/s^2}$	Steering wheel angle – lateral acceleration gradient at a lateral acceleration of 4 m/s^2.
Sideslip angle	$\left.\frac{\partial \beta}{\partial \delta_H}\right\|_{4\ m/s^2}$	Sideslip angle – steering wheel angle gradient at a lateral acceleration of 4 m/s^2.
	$\left.\frac{\partial \beta}{\partial a_y}\right\|_{4\ m/s^2}$	Sideslip angle – lateral acceleration gradient at a lateral acceleration of 4 m/s^2.
Maximal grip	$a_{y,max}$	Maximum lateral acceleration.

Concerning the ramp steer, the possible OMC describe the two main characteristics of the steady-state vehicle behaviour (Niedermeier, Peckelsen, & Gauterin, 2013): the characteristics of steering angle over lateral acceleration and the characteristics of sideslip angle over lateral acceleration. The OMC are enlisted in Table 5.14.

Concerning the three transient manoeuvres (continuous sine sweep, linearity and lane change) the possible OMC are derived from gain function, delay function and overshoot function of the VDF as presented in section 5.1.2.1. Moreover, known OMC are taken into considerations and influence the definition of the following possible OMC (see section 2.2.1). The possible OMC for the continuous sine sweep are enlisted in Table 5.15: they described the strength of the vehicle reaction, its variation over steering wheel angle and the time delays. The time delays are defined to describe the relative time delays between the three vehicle reaction to a steering input: first, the yaw velocity rises, then, the lateral acceleration increases and, finally, the roll angle is built up.

Table 5.15: Possible OMC for continuous sine sweep

Characteristics	Formulation	Description
Strength of reaction	$\max\left(\frac{X}{\delta_H}\right)$	Maximal gain of roll angle, sideslip angle and lateral acceleration. $X \in [\varphi, \beta, a_y]$
Influence of steering wheel amplitude	$\frac{X_{\delta_{H,max}} - X_{\delta_{H,min}}}{\delta_{H,max} - \delta_{H,min}}$	Increment of the gain of roll angle, sideslip angle and lateral acceleration. $X \in [\varphi, \beta, a_y]$
Time delay	$\max(\Delta t_{\dot{\psi}-\delta_H})$ $\max\left(\Delta t_{a_y-\dot{\psi}}\right)$ $\max\left(\Delta t_{\varphi-a_y}\right)$	Time delay of yaw velocity, lateral acceleration and roll angle.

Table 5.16: Possible OMC for linearity and lane change

Characteristics	Formulation	Description
Strength of reaction	$\left.\frac{X}{\delta_H}\right\|_{ss}$	Gain of roll angle, sideslip angle and lateral acceleration at steady-state ("ss") conditions. $X \in [\varphi, \beta, a_y]$
Time delay	$\max(\Delta t_{\dot{\psi}-\delta_H})$ $\max(\Delta t_{a_y-\dot{\psi}})$ $\max(\Delta t_{\varphi-a_y})$	Time delay of yaw velocity, lateral acceleration and roll angle.
Overshoot	$\frac{X_{max} - X_{ss}}{X_{ss}}$	Overshoot peak of roll angle, sideslip angle and lateral acceleration in relation to steady-state value ("ss"). $X \in [\varphi, \beta, a_y]$

The possible OMC for linearity and lane change are enlisted in Table 5.16: they described the vehicle reaction strength, the time delays and the overshoot peaks for two different step steering inputs and, subsequently, two different lateral accelerations.

5.1.3.2 Objectivation Study at the Dynamic Driving Simulator

The objectivation study at the dynamic driving simulator focuses on the transient manoeuvres, i.e. the continuous sine sweep, the linearity and the lane change as described in section 5.1.1. The choice of the relevant OMC for the ramp steer is already motivated in (Niedermeier, Peckelsen, & Gauterin, 2013). The objectivation study aims to asses which objective criteria describe best a subjective feeling.

The objectivation study is based on a DoE composed of 25 virtual tyres, generated through the Tyre Shaper (see section 3.3.1). The number of tyres is limited by the available time at the driving simulator; more tyres would increase the statistical significance of the objectivation. The method for generating the DoE is chosen according to the homogeneity index (see section 3.4.2.4): the most homogeneous design for the present DoE is generated by a latin hypercube. All tyres are mounted on three different vehicles: a compact vehicle, a middle-class SUV and a high-class limousine. Subsequently, 75 realistic tyre-vehicle combinations are generated (see Table 5.17). For each vehicle a reference tyre-vehicle combination is added. The simulations are performed using the MF model 5.2 (see section 3.2.2.1) and an improved double-track vehicle model.

The simulations are evaluated by the test drivers in the dynamic driving simulator (Reichelt & Strackerjan, 1992). Four test drivers participate to the objectivation study: one for each vehicle and a fourth for the high-class limousine. Subsequently, the tyres of the high-class limousine are evaluated twice: it is shown, that the SMI of the two test drivers of the high-class limousine are perfectly comparable.

Table 5.17: Objectivation study at the dynamic driving simulator

Parameter	Number	Comment
Tyres	$25 + 1$	Virtual tyres generated through the Tyre Shaper, plus a reference tyre.
Vehicles	3	A compact vehicle, a SUV and a high-class limousine.
Manoeuvres	3	Continuous sine sweep, linearity, lane change.
Test Drivers	4	One for each vehicle and a fourth for the high-class limousine.

Figure 5.18: BMW dynamic driving simulator

The dynamic driving simulator, represented in Figure 5.18, is composed of a carbon-fibre sphere mounted on a hexapod. The carbon-fibre sphere contains a real BMW vehicle. The steering wheel of the vehicle is not moved by the test driver, but by actuators according to the simulation signals. Of course, the diver keeps his hands on the steering wheel in order to feel the steering input. An optimisation algorithm (Sammet, 2007) calculates the movements of the actuators of the simulator in advanced (open-loop manoeuvre), guaranteeing an exact reproduction of the dynamics without introducing scaling factors and without adding time delays: moreover, the optimisation parameters are set to reduce to a minimum the tilting mechanisms used to generate lateral acceleration, and, subsequently, their influence on the subjective evaluation.

Although test drivers are trained to evaluate as objectively as possible, a subjective influence on the experienced vehicle dynamics cannot be avoid (Wolf, 2009). The detection thresholds for yaw velocity, yaw acceleration and lateral acceleration are analysed in (Tomaske, 1983), (Youngblut, Johnston, Nash, Wienclaw, & Will, 1996) and (VDI 2057 Blatt 1, 2002). In order to reduce to a minimum the subjective influence, all 25 virtual tyres are evaluated against a reference tyre. Moreover, each test driver undergoes a learning phase (15 minutes, different tyres, and different manoeuvres). After the learning phase, test drivers are familiar with the virtual environment and have adapted their subjective evaluation to the open-loop manoeuvres. For example, they have memorised the steering wheel movement and are able to evaluate the dynamics as if they would start the steering wheel input.

In order to quantify the plausibility of the subjective evaluations at the dynamic driving simulator, a comparison between test track and simulator is performed. The lateral dynamics of two real tyres is evaluated on a compact sport vehicle by three different test drivers for the linearity and for the lane change. The first tyre is a standard R18 high-performance tyre; the second is a standard R18 comfort tyre. Table 5.19 shows the results: the average SMI on the test track are comparable to those at the dynamic driving simulator. Not only the SMI, but also the annotations of the test drivers are very similar.

Table 5.19: Case study, comparison between average SMI on the test track and at the dynamic driving simulator

	Tyre 1		Tyre 2	
	Test track	Simulator	Test track	Simulator
Linearity	7.58	7.58	7.64	7.58
	Slightly progressive, especially at the end of the manoeuvre.	Slightly progressive, especially at the end of the manoeuvre.	More linear than tyre 1.	Very similar to tyre 1, maybe more linear.
Lane change	7.83	7.75	7.50	7.42
	Direct and sporty reaction, very fast.	Very sporty, the vehicle is very stable.	Almost as stable as tyre 1, but slightly slower.	More roll angle; slower and weaker reaction.

5.1.3.3 Choice of Objective Manoeuvre Criteria

The choice of the relevant OMC for each manoeuvre is done according to a global sensitivity analysis (GSA) based on the 75 virtual and the additional three reference tyre-vehicle combinations used for the objectivation study at the dynamic driving simulator.

The design variables are the OMC of each tyre-vehicle combination; the design targets are the relative SMI. According to Sobol's GSA (see section 3.5.2.1), the higher is Sobol's total index "T", the stronger is the influence of the OMC on the SMI.

Table 5.20: GSA results concerning the OMC of continuous sine sweep

	SMI ($R^2 = 0.90$)	
	M	T
$\max\left(\frac{\varphi}{\delta_H}\right)$	0.06	**0.11**
$\max\left(\frac{\beta}{\delta_H}\right)$	0.14	0.09
$\max\left(\frac{a_y}{\delta_H}\right)$	0.04	0.04
$\frac{\partial\varphi}{\partial\delta_H}$	0.29	0.08
$\frac{\partial\beta}{\partial\delta_H}$	0.02	0.10
$\frac{\partial a_y}{\partial\delta_H}$	0.03	0.06
$\max(\Delta t_{\dot{\psi}-\delta_H})$	0.17	**0.12**
$\max\left(\Delta t_{a_y-\dot{\psi}}\right)$	0.39	**0.30**
$\max\left(\Delta t_{\varphi-a_y}\right)$	0.11	0.09

Moreover, the R^2-values of the artificial neuronal networks (ANN) needed to determine Sobol's indices are calculated. As some R^2-values are below 0.95, the results are checked graphically (see section 3.5.2.2) due to the high complexity of the correlations and the relatively limited number of experiments. For the sake of completeness also Sobol's main index "M" is reported.

Concerning the continuous sine sweep, the results of the GSA are shown in Table 5.20. Time delays concerning yaw velocity and lateral acceleration, as well as roll angle gain show the highest sensitivities. A vehicle is evaluated good if it rolls less and responses directly to the steering input.

Figure 5.21 shows exemplarily the graphical relation between SMI and the time delay of lateral acceleration of all 78 tyre-vehicle combinations. It should be noted, that several tyres are grouped in the upper-left part of the diagram: the time delay of these tyres is below the average and evaluated as good. Only some tyres show high time delays; these tyres are evaluated with low SMI. Concluding, high time delays stand out as bad; low time delay are good, but the subjective evaluation tends to saturate.

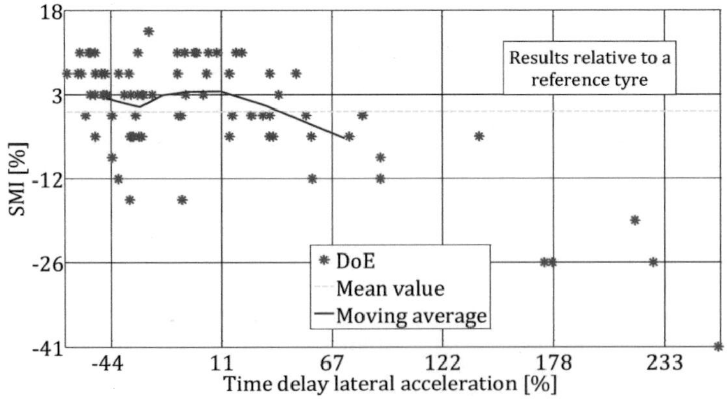

Figure 5.21: Graphical analysis for continuous sine sweep

Table 5.22: GSA results concerning the OMC of linearity

	SMI ($R^2 = 0.82$)	
	M	T
$\left.\frac{\varphi}{\delta_H}\right\|_{ss}$	0.02	0.05
$\left.\frac{\beta}{\delta_H}\right\|_{ss}$	0.20	**0.26**
$\left.\frac{a_y}{\delta_H}\right\|_{ss}$	0.06	0.08
$\max(\Delta t_{\dot{\psi}-\delta_H})$	0.16	**0.14**
$\max\left(\Delta t_{a_y-\dot{\psi}}\right)$	0.18	**0.10**
$\max\left(\Delta t_{\varphi-a_y}\right)$	0.06	0.05
$\frac{\varphi_{max}-\varphi_{ss}}{\varphi_{ss}}$	0.03	0.08
$\frac{\beta_{max}-\beta_{ss}}{\beta_{ss}}$	0.04	0.09
$\frac{a_{y,max}-a_{y,ss}}{a_{y,ss}}$	0.06	**0.15**

Concerning linearity, the results of the GSA are shown in Table 5.22. Sideslip angle gain, time delays concerning yaw velocity and lateral acceleration, as well as lateral acceleration overshoot show the highest sensitivities. A low sideslip angle gain combined with a low lateral acceleration overshoot characterizes a strong and stable vehicle response; low time delays indicate a direct vehicle reaction to the steering input. Figure 5.23 shows exemplarily the graphical relation between SMI and the sideslip angle gain. Although the correlation is poor, it is possible to recognise a trend: the higher the slip angle gain, the worse the SMI. However, for low slip angle gains, the trend saturates.

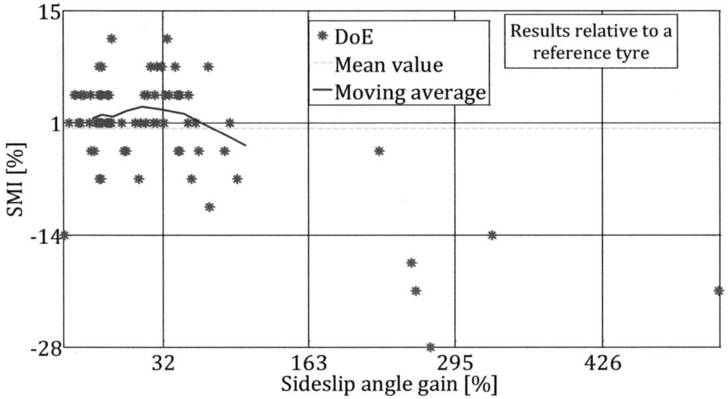

Figure 5.23: Graphical analysis for linearity

Concerning lane change, the results of the GSA are shown in Table 5.24. Time delays concerning yaw velocity and roll angle, as well as sideslip angle gain and overshoot show the highest sensitivities. The results are similar to those of the linearity: low time delays indicate a direct vehicle reaction, low sideslip angle a stable reaction. Moreover, roll angle gains importance: in fact, at higher lateral accelerations the time delay between roll angle and lateral acceleration increases, generating a less harmonic vehicle response. Therefore, low values of roll angle delay become crucial.

Table 5.24: GSA results concerning the OMC of lane change

	SMI ($R^2 = 0.91$)	
	M	T
$\left.\frac{\varphi}{\delta_H}\right\|_{ss}$	0.05	0.02
$\left.\frac{\beta}{\delta_H}\right\|_{ss}$	0.16	**0.14**
$\left.\frac{a_y}{\delta_H}\right\|_{ss}$	0.13	0.07

$\max(\Delta t_{\dot{\psi}-\delta_H})$	0.31	**0.32**
$\max\left(\Delta t_{a_y-\dot{\psi}}\right)$	0.20	0.08
$\max\left(\Delta t_{\varphi-a_y}\right)$	0.05	**0.14**
$\frac{\varphi_{max}-\varphi_{ss}}{\varphi_{ss}}$	0.01	0.08
$\frac{\beta_{max}-\beta_{ss}}{\beta_{ss}}$	0.08	**0.10**
$\frac{a_{y,max}-a_{y,ss}}{a_{y,ss}}$	0.12	0.05

Figure 5.25 and Figure 5.26 show exemplarily the graphical relation between SMI and the sideslip angle gain as well as the time delay of yaw velocity. Both diagrams show a clear relation characterized by a saturation segment in the high SMI range and an almost parabolic-to-linear decrement of the SMI for higher slip angle gains and time delays.

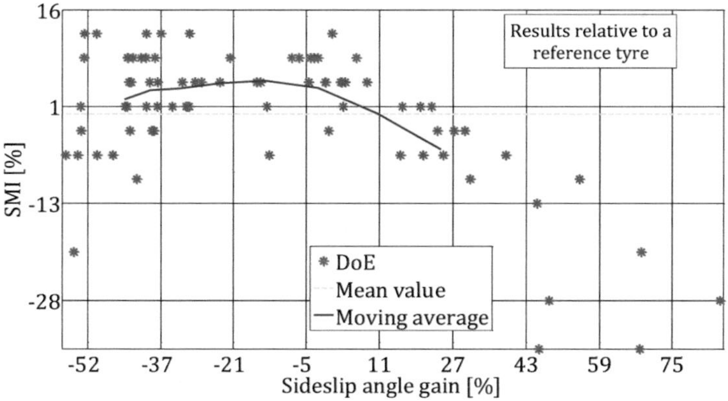

Figure 5.25: Graphical analysis for lane change

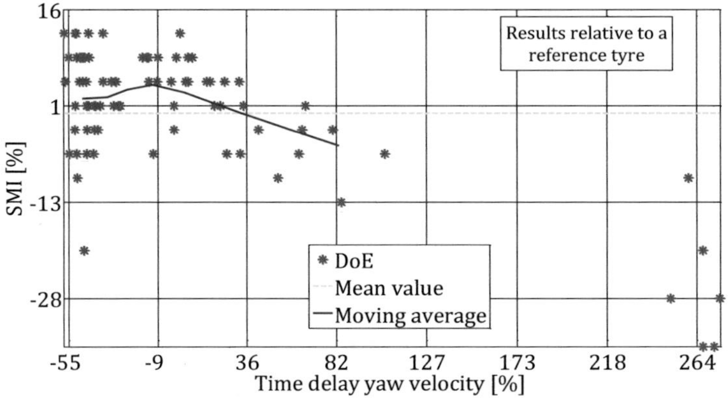

Figure 5.26: Graphical analysis for lane change

It should be noticed, that in the lane change the values of the OMC are covered more homogeneously than in the linearity and in the continuous sine sweep: the more severe driving conditions of the lane change (higher lateral acceleration) increase the differences between the artificial tyres.

Table 5.27: Relevant OMC for lateral dynamics

Ramp steer	Continuous sine sweep	Linearity	Lane change
$\delta_H\|_{4\ m/s^2}$	$\max\left(\frac{\varphi}{\delta_H}\right)$	$\frac{\beta}{\delta_H}\Big\|_{ss}$	$\frac{\beta}{\delta_H}\Big\|_{ss}$
$\frac{\partial \delta_H}{\partial a_y}\Big\|_{4\ m/s^2}$	$\max(\Delta t_{\dot{\psi}-\delta_H})$	$\max(\Delta t_{\dot{\psi}-\delta_H})$	$\max(\Delta t_{\dot{\psi}-\delta_H})$
$\frac{\partial \beta}{\partial \delta_H}\Big\|_{4\ m/s^2}$	$\max\left(\Delta t_{a_y-\dot{\psi}}\right)$	$\max\left(\Delta t_{a_y-\dot{\psi}}\right)$	$\max\left(\Delta t_{\varphi-a_y}\right)$
$\frac{\partial \beta}{\partial a_y}\Big\|_{4\ m/s^2}$		$\frac{a_{y,max}-a_{y,ss}}{a_{y,ss}}$	$\frac{\beta_{max}-\beta_{ss}}{\beta_{ss}}$

In Table 5.27 the relevant OMC for the chosen manoeuvres of the lateral dynamics are enlisted (The OMC for the ramp steer are discussed in section 5.1.3.1). The time delay of yaw velocity, characterising the vehicle response time, is always evaluated as important. Sideslip angle and roll angle become crucial at higher lateral accelerations: in fact, stability gains importance with lateral acceleration.

5.2 Functional Tyre Characteristics

In this section, first, the possible functional tyre characteristics (FTCs) describing the relevant characteristics of the tyre lateral dynamics are defined; then, their influence on the OMC is analysed and the relevant FTCs are chosen.

5.2.1 Definition of Functional Tyre Characteristics

The FTCs presented in this section should allow a unique and exhaustive description of the tyre lateral dynamics. The choice is based on (Niedermeier, Peckelsen, & Gauterin, 2013). However, the definition of three FTCs changes, namely of the "maximal lateral friction" (here defined as friction coefficient and not as maximal lateral force), of the "position of maximal lateral friction" (here defined as percentage variation and not as absolute value) and of the "degression of lateral force over slip angle" (here defined in relation to $\alpha_{y,max}$ and not to an absolute value) (compare Table 2.9 and Table 5.28). Moreover, the variation due to camber angle, tyre load and longitudinal slip is defined as percentage variation. These changes allow to uncouple better the FTCs.

As described in section 2.2.2.1 the set of FTCs is composed of five main FTCs describing the lateral force over slip angle (see Table 5.28).

Table 5.28: The five main FTCs for lateral dynamics (defined at a given tyre load $F_{z,ref}$ and inflation pressure p)

Name	Formulation	Symbol	Unit
Lateral force at zero slip angle	$F_y\|_{0°}$	$F_{y,0}$	$[N]$
Cornering stiffness	$\left.\frac{\delta F_y}{\delta \alpha}\right\|_{0°}$	$K_{y,0}$	$[N/°]$
Maximal lateral friction	$\max\left(\frac{F_y}{F_z}\right)$	$\mu_{y,max}$	$[N/N]$
Position of maximal lateral friction	$\frac{\alpha_{y,max} - \frac{max(F_y) - F_y\|_0}{K_{y,0}}}{\frac{max(F_y) - F_y\|_0}{K_{y,0}}} \cdot 100$	$\alpha_{max,\%}$	$[\%]$
Degression of lateral force over slip angle	$\frac{\left(\mu_y\|_{1.25\alpha_{y,max}} - \mu_{y,max}\right)/\mu_{y,max}}{0.25\alpha_{y,max}} \cdot 100$	$\mu'_{y,max,\alpha}$	$[\%/°]$

Additional 15 FTCs describe the linear percentage variation of the five main FTCs due to camber angle $[\%/°]$, tyre load $[\%/N]$ and longituidinal slip $[\%/-]$. For example, the variation of the cornering stiffness K_y due to tyre load is:

$$K'_{y,0,F_z} = \frac{\Delta K_y/|K_{y,0}|}{\Delta F_z} \cdot 100 = \frac{\left(K_y|_{F_{z,max}} - K_y|_{F_{z,min}}\right)/|K_{y,0}|}{F_{z,max} - F_{z,min}} \cdot 100 \quad \textbf{5.10}$$

where K_y and F_z are respectively the cornering stiffness and the tyre load, and $K_{y,0}$ is the cornering stiffness at reference tyre load.

Table 5.29: Ranges for the calculation of the FTC variations for lateral dynamics

Variation	Symbol	Range
Camber angle	γ	$[-4°, 0°]$
Tyre load	F_z	$[F_{z,ref} - 1500N, F_{z,ref} + 1500N]$
Longitudinal slip	κ	$[-0.1, 0.1]$

The deltas are calculated between a reference minimum and a reference maximum tyre load. The variation ranges for tyre load, camber angle and longitudinal slip are defined in Table 5.29: they are chosen to generate a realistic variation for normal driving conditions ($F_{z,ref}$ represents the average tyre load for a given vehicle). The range of camber angle is asymmetric, as the MF model 5.2 is asymmetric and as the camber angle is always chosen negative (in order to guarantee a stable straight-forward driving). Two additional FTCs describe the lateral relaxation length and its variation due to tyre load (see Table 5.30).

Table 5.30: Two additional FTCs for lateral dynamics (defined at a given tyre load $F_{z,ref}$ and inflation pressure p)

Name	Formulation	Symbol	Unit
Lateral relaxation length	$\sigma_y\|_0$	$\sigma_{y,0}$	$[m]$
Variation of lateral relaxation length due to tyre load	$\frac{\Delta\sigma_y/\|\sigma_{y,0}\|}{\Delta F_z} \cdot 100$	$\sigma'_{y,0,F_z}$	$[\%/N]$

5.2.2 Choice of Functional Tyre Characteristics

The choice of the relevant FTCs is done according to a global sensitivity analysis (GSA) based on the same design of experiments (DoE) used for the objectivation study consisting of the 78 tyre-vehicle combinations. The chosen manoeuvres are performed using the MF model 5.2 for the tyres (see section 3.2.2.1) and a double-track model for the vehicle.

The experiments of the DoE are virtual tyres generated through the Tyre Shaper (see section 3.3.1). The FTC variations are set to cover the ranges measured on real tyres. The design variables are the 22 FTCs for lateral dynamics; the design targets are the 15 OMC. According to Sobol's GSA (see section 3.5.2.1), the higher is Sobol's total index "T", the stronger is the influence of the FTCs on the OMC. Moreover, the R^2-values of the artificial neuronal networks (ANN) needed to determine Sobol's are calculated.

The results of the analysis are shown in Table 5.31: for each manoeuvre the five dominant FTCs are enlisted. They are ranked by importance according to Sobol's total sensitivity index. The OMC regarding the ramp steer is taken from (Niedermeier, Peckelsen, & Gauterin, 2013). In the steady-state manoeuvres the maximal lateral friction $\mu_{y,max}$ dominates. For transient manoeuvres cornering stiffness $K_{y,0}$ is evaluated as the most important FTC. Moreover, for the lane change manoeuvre also the lateral relaxation length is relevant: in fact, the vehicle response is determined by the cornering stiffness (i.e. the cornering stiffness and the lateral force at zero slip angle, both generating a pre-tension of the vehicle axle) and the time needed to build up the lateral force (i.e. the lateral relaxation length). It should be noted, that maximal lateral friction is not evaluated as important, as in transient manoeuvres the focus lies on the vehicle reaction time and its stability, which is given mainly by the cornering stiffnesses and their variation due to camber and load.

Table 5.31: Relevant FTCs for lateral dynamics subdivided by manoeuvre

Rank	Ramp steer	Continuous sine sweep	Linearity	Lane change
1	$\mu_{y,max}$	$K_{y,0}$	$K_{y,0}$	$K_{y,0}$
2	$K_{y,0}$	$K'_{y,0,F_z}$	$\mu'_{y,max,\gamma}$	$F_{y,0}$
3	μ'_{y,max,F_z}	$\alpha_{max,\%}$	$K'_{y,0,F_z}$	$\sigma_{y,0}$
4	$K'_{y,0,F_z}$	μ'_{y,max,F_z}	μ'_{y,max,F_z}	$K'_{y,0,F_z}$
5	$K'_{y,0,\gamma}$	$F_{y,0}$	$K'_{y,0,\gamma}$	μ'_{y,max,F_z}

Finally, five FTCs are chosen as most representative for all four manoeuvres of the lateral dynamics. The choice is done taking into account the ranking presented in Table 5.31 and the possibility to have an unique description of the complete curve of the lateral force. The chosen FTCs are enlisted in Table 5.32. The first two FTCs are among the most important for lateral dynamics: maximal lateral friction $\mu_{y,max}$ is ranked first for ramp steer; the cornering stiffness $K_{y,0}$ is the most sensitive for continuous sine sweep, linearity and lane change (see Table 5.31). The position of maximal lateral friction $\alpha_{max,\%}$ is added to allow a complete description of the lateral force characteristics as a function of slip angle. The last two FTCs characterize the two relevant variations of cornering stiffness and maximal friction coefficient due to tyre load; both are among the most five sensitive FTCs according to Table 5.31.

Table 5.32: Relevant FTCs for lateral dynamics (defined at a given tyre load $F_{z,ref}$ and inflation pressure p)

Name	Formulation	Symbol	Unit
Cornering stiffness at zero slip angle	$\left.\frac{\delta F_y}{\delta \alpha}\right\|_0$	$K_{y,0}$	$[N/°]$
Maximal lateral friction	$\max\left(\frac{F_y}{F_z}\right)$	$\mu_{y,max}$	$[N/N]$
Position of maximal lateral friction	$\frac{\alpha_{y,max} - \frac{max(F_y) - F_y\|_0}{K_{y,0}}}{\frac{\mu_{y,max} - F_y\|_0}{K_{y,0}}} \cdot 100$	$\alpha_{max,\%}$	$[\%]$
Variation of cornering stiffness due to tyre load	$\frac{\Delta K_y / \|K_{y,0}\|}{\Delta F_z} \cdot 100$	$K'_{y,0,F_z}$	$[\%/N]$
Variation of maximal lateral friction due to tyre load	$\frac{\Delta \mu_y / \|\mu_{y,max}\|}{\Delta F_z} \cdot 100$	μ'_{y,max,F_z}	$[\%/N]$

5.3 Summary

In section 1 the requirement lateral dynamics is analysed. First, the choice of manoeuvres is motivated. Four manoeuvres are chosen: the "ramp steer" (see Table 5.1) to characterize the steady-state behaviour up to maximal lateral acceleration, the "continuous sine sweep" (see Table 5.2) and the "linearity" (see Table 5.4) to evaluate the vehicle reaction to a sinusoidal and to a step steering input, and the "lane change" (see Table 5.6) to judge the vehicle response and stability at higher lateral acceleration.

Then, the "Vehicle Dynamics Formulae" (VDF) are introduced (see section 5.1.2.1) to define a unique set of possible objective manoeuvre criteria

(OMC): knowing the relevant VDF parameters for reproducing a given vehicle response (e.g. the lateral dynamics in a lane change), allows to derive a non-redundant set of possible OMC. After having defined the possible OMC (see Table 5.15 and Table 5.16), 15 relevant OMC are identified through GSA based on the objectivation study at the dynamic driving simulator (see Table 5.27).

Finally, 22 functional tyre characteristics (FTCs) concerning lateral dynamics are introduced and a second global sensitivity analysis (GSA) is performed. The analyses are based on the same 78 tyre-vehicle combinations of the objectivation study at the dynamic driving simulator: five FTCs are chosen as relevant for describing the lateral dynamics (see Table 5.32).

6 Ride Comfort

6.1 Objective Manoeuvre Criteria

In this section, first, the manoeuvres used to evaluate ride comfort of vehicles are presented; then, the objective manoeuvre criteria (OMC) necessary to describe the ride comfort characteristics are defined.

6.1.1 Choice of Manoeuvres

The manoeuvres for ride comfort are defined to excite best the relevant vertical dynamics of vehicle and wheel-suspension system. According to considerations of section 2.3.2 the frequency range for ride comfort can be limited to 30 Hz: to identify the characteristics of the spectra of the vehicle response, harmonic excitations are suggested (see section 0). The road sweep manoeuvre is chosen based on considerations presented in (Di Luise, 2015).

6.1.1.1 Road Sweep

The road sweep is chosen as it represents the best compromise between accuracy and evaluation time. It is performed at constant driving velocity v driving straight-forward (see Table 6.1). The road irregularity z_e is defined by a sinusoidal function:

$$z_e(x) = \bar{z}_e \, sin\left(2\pi f_e \frac{x}{v}\right) \qquad 6.1$$

where amplitude $\bar{z}_e$ and frequency f_e change quasi-statically along the longitudinal displacement x (see Figure 6.2). It first excites the front axle (same excitation on left and right tyre) and then the rear axle, where the time delay is given by the ratio of vehicle wheel base and driving velocity.

Table 6.1: Road sweep, manoeuvre parameters

Name	Symbol	Value / Range
Driving velocity	v	80 km/h
Road excitation amplitude	$\bar{z}_e$	Quasi-statically decreasing from 0.02 m to 0 m (quadratically).
Road excitation frequency	f_e	Quasi-statically increasing from 0.5 Hz to 30 Hz (linearly).

In Figure 6.2 and Figure 6.3 the road irregularity is shown in the space domain and in the frequency domain. Excitation amplitude and frequency are chosen to best reproduce realistic irregularities that may occur on country roads.

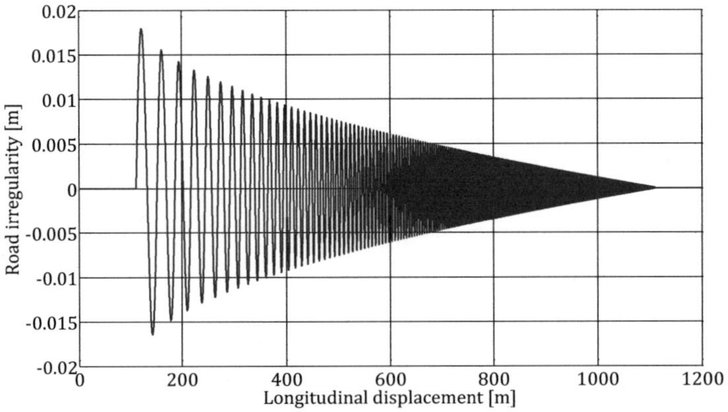

Figure 6.2: Road sweep, road irregularity in the space domain

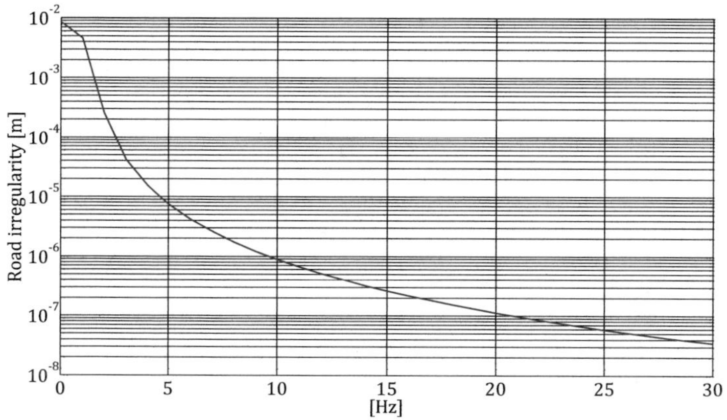

Figure 6.3: Road sweep, road irregularity in the frequency domain

6.1.2 Choice of Objective Manoeuvre Criteria

In this section, the set of relevant objective manoeuvre criteria (OMC) is presented. The choice is based on considerations presented in (Guiggiani, 2014) and explained in (Di Luise, 2015). The selected OMC describe the spectra of vehicle pitching $\ddot{\vartheta}$ and bouncing acceleration $a_{z,s}$ and of wheel hub vertical acceleration $a_{z,u}$. All accelerations are evaluated either in the centre of mass of the vehicle or at the wheel hub.

In order to transform the acceleration signals from the time domain to the frequency domain the fast Fourier transform (FFT) is used. This algorithm, implemented in Matlab's FFT library (Frigo & Johnson, 1998), computes the discrete Fourier transform (DFT) using, amongst others, the Cooley-Tukey algorithm (Cooley & Tukey, 1965). In order to reduce the variance in the spectra, Welch's method (Welch, 1967) is applied: the spectra are generated making a time-average of the periodograms of overlapping time segments of the sweep manoeuvre. This causes that, the magnitudes of the accelerations are average values (the excitation is not periodic) and the frequency resolution is the inverse of the time segment length: in our case the time segments last 2 s and, subsequently, the frequency resolution is $\Delta f = 0.5\ Hz$.

Concerning the OMC of the vehicle, the focus lies on the pitching $\ddot{\vartheta}$ and bouncing a_z acceleration. In Figure 6.4 the spectrum of the bouncing acceleration is exemplarily shown for two different vehicles. Both vehicles have their centre of mass approximately in the middle of the wheel base. Trajectory velocity and road excitation are set as described in section 6.1.1.1 in order to excite successively all frequencies between 0 and 30 Hz. In a spectrum measured at the vehicle centre of mass the peaks are generated through the superposition of the eigenmodes of the vehicle centre of mass and of the wheel-suspension system of front and rear axle, which are, moreover, subjected to driving velocity and vehicle wheel base dependent time-delay. Subsequently, it is not possible to link directly peaks of the spectrum to eigenmodes of the vehicle: generally, the first peak is caused by the vehicle body and the following by the unsprung masses. It could be noticed that the compact sport vehicle has lower mass and shorter wheel base causing the frequencies to increase.

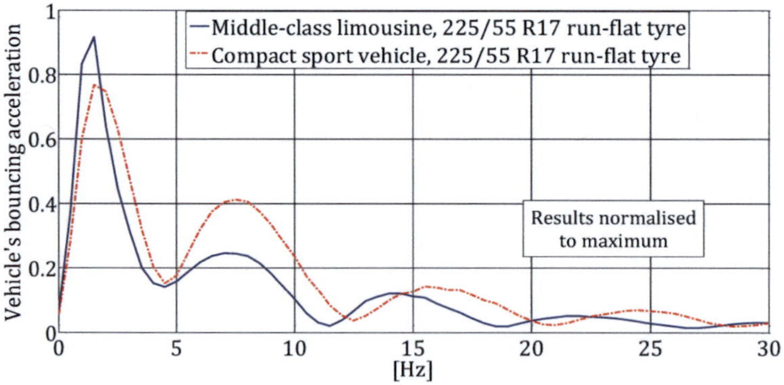

Figure 6.4: Spectrum of the vertical acceleration evaluated in the vehicle centre of mass (simulated with a double-track and the MF-SWIFT model 6.1.2)

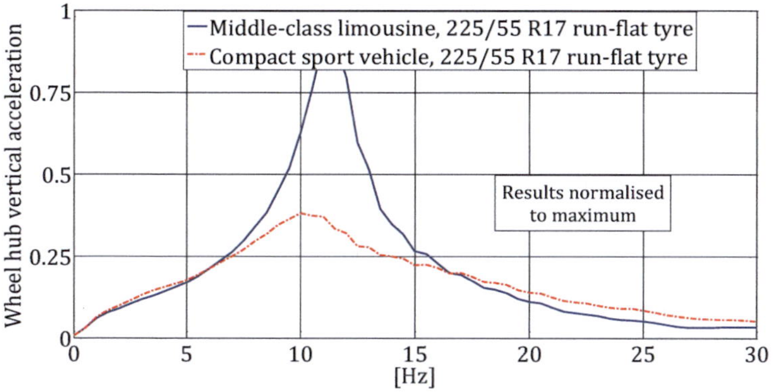

Figure 6.5: Spectrum of the vertical acceleration of the front axle wheel hub (simulated with a double-track and the MF-SWIFT model 6.1.2)

Concerning the OMC of the wheel hub, the focus lies on the vertical acceleration of front $a_{z,front}$ and rear $a_{z,rear}$ axle. In Figure 6.5 the spectrum of the vertical acceleration of the front axle is exemplarily shown for two different vehicles: the eigenfrequency of the wheel-suspension system (7 to 15 Hz) is clearly visible. As the compact sport vehicle has higher damping elements, also the peak value is lower.

For both, bouncing and pitching acceleration, three OMC are defined (see Table 6.6): they describe the peak value at low frequencies and the area below it, as well as the total area in the frequency range between 0 and 30 Hz. Concerning the OMC of the wheel hub for front and rear axle, three OMC are defined: they describe the peak value and the frequency of the wheel hub resonance, as well as the total area in the frequency range between 0 and 30 Hz. Pitching and bouncing accelerations are measured in the vehicle mass centre (see Table 6.6).

Table 6.6: Relevant OMC for ride comfort

Characteristics	Formulation	Description
Vehicle $X \in [\ddot{\vartheta}, a_z]$	$\max(\lvert X \rvert)$	Peak value at low frequencies.
	$\int_{0.5}^{3} \lvert X \rvert df$	Area below the first peak at low frequencies.
	$\int_{0}^{30} \lvert X \rvert df$	Total area in the frequency range of ride comfort.
Wheel hub $i \in [front, rear]$	$\max(\lvert a_{z,u_i} \rvert)$	Peak value of the resonance.
	$f_{a_{z,u_i}}$	Frequency of the resonance.
	$\int_{0}^{30} \lvert a_{z,u_i} \rvert df$	Total area in the frequency range of ride comfort.

6.2 Functional Tyre Characteristics

In this section, first, the functional tyre characteristics (FTCs) describing the relevant characteristics of the tyre vertical dynamics are defined; then, their influence on the objective manoeuvre criteria (OMC) is analysed and the relevant FTCs are chosen. Some preliminary analyses are presented in (Di Luise, 2015).

6.2.1 Definition of Functional Tyre Characteristics

The FTCs presented in this section should allow a unique and exhaustive description of the tyre vertical dynamics.

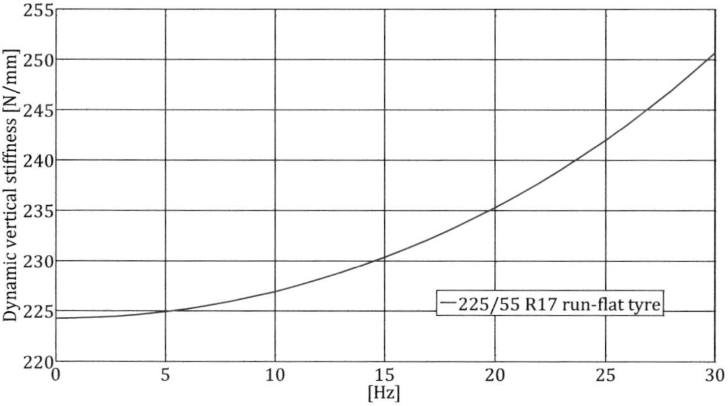

Figure 6.7: Dynamic vertical stiffness (simulated with the MF-SWIFT model 6.1.2)

The vertical dynamics of the vehicle is influenced by the dynamic vertical stiffness of the tyre, which is almost parabolically increasing between 0 and 30 Hz (see exemplarily Figure 6.7) and has zero slope at 0 Hz. Subsequently, two FTCs are defined to describe its characteristics (see Table 6.8).

Table 6.8: The two main FTCs for ride comfort (defined at a given tyre load $F_{z,ref}$ and inflation pressure p)

Name	Formulation	Symbol	Unit
Dynamic vertical stiffness at zero hertz	$K_z\|_0$	$K_{z,0}$	$[N/m]$
Slope of dynamic vertical stiffness	$\left.\frac{\partial K_z}{\partial f}\right\|_{14\,Hz}$	$K'_{z,14,f}$	$[Ns/m]$

Table 6.9: Ranges for the calculation of the FTC variations for ride comfort

Variation	Symbol	Range
Tyre load	F_z	$[0.75F_{z,ref}, 1.25F_{z,ref}]$
Trajectory velocity	v_x	$[0.75v_{x,ref}, 1.25v_{x,ref}]$

Four additional FTCs describe the linear percentage variation of the two main FTCs due to tyre load $[\%/N]$ and trajectory velocity $[\%/m/s]$. For example, the variation of the dynamic vertical stiffness K_z due to tyre load F_z is:

$$
\begin{aligned}
K'_{z,0,F_z} &= \frac{\Delta K_z/|K_{z,0}|}{\Delta F_z} \cdot 100 \\
&= \frac{\left(K_z|_{F_{z,max}} - K_z|_{F_{z,min}}\right)/|K_{z,0}|}{F_{z,max} - F_{z,min}} \cdot 100
\end{aligned}
\qquad 6.2
$$

where K_z and F_z are respectively the dynamic vertical stiffness and the tyre load, and $K_{z,0}$ is the dynamic vertical stiffness at zero hertz. The deltas are calculated between a reference minimum and a reference maximum tyre load. The variation ranges for tyre load and trajectory velocity are defined in Table 6.9: they are chosen to generate a realistic variation for normal driving conditions. The reference tyre load $F_{z,ref}$ represents the average tyre load for a given vehicle, while the reference trajectory velocity $v_{x,ref}$ is the trajectory velocity of the road sweep (in our case 80 km/h).

6.2.2 Choice of Functional Tyre Characteristics

The choice of the relevant FTCs is done according to a graphical sensitivity analysis (see section 3.5.2.2) based on a design of experiment (DoE) consisting of 75 virtual tyres. The chosen manoeuvres are performed using the MF-SWIFT model 6.1.2 for the tyres (see section 3.2.2.2) and a double-track model for the vehicle.

The experiments of the DoE are virtual tyres generated through the Tyre Shaper (see section 3.3.1). The FTC variations are set to cover the ranges measured on real tyres. The design variables of the DoE are the six FTCs for ride comfort; the design targets the twelve OMC. The method for generating the DoE is chosen according to the homogeneity index (see section 3.4.2.4): the most homogeneous design for six design variables and 75 experiments is generated by a latin hypercube. The tyres are mounted on a middle-class limousine. The analysis is repeated for a second vehicle, a compact sport vehicle: the results are comparable, the choice of the relevant FTCs is independent from the vehicle. The selection of the most sensitive FTCs is based on graphical and correlation analyses presented in (Di Luise, 2015). The most important results are presented in this section.

The dynamic vertical stiffness is the most important FTC. The other FTCs, especially the influence of tyre load and trajectory velocity, are negligible. This is also proven by analytical considerations in (Di Luise, 2015). Concerning the vehicle movements, the increase of the dynamic vertical stiffness causes on the one hand a decrement of the peak value and of its area at low frequencies, on the other hand an increment of the total area in the frequency range of ride comfort. The results concerning these two effects are presented in Figure 6.10 and Figure 6.11.

Concerning the wheel hub (see Figure 6.12), if the dynamic vertical stiffness increases all OMC of the wheel hub increase (for front and rear axle), as a stiffer tyre filters less the road irregularities causing a worse ride comfort.

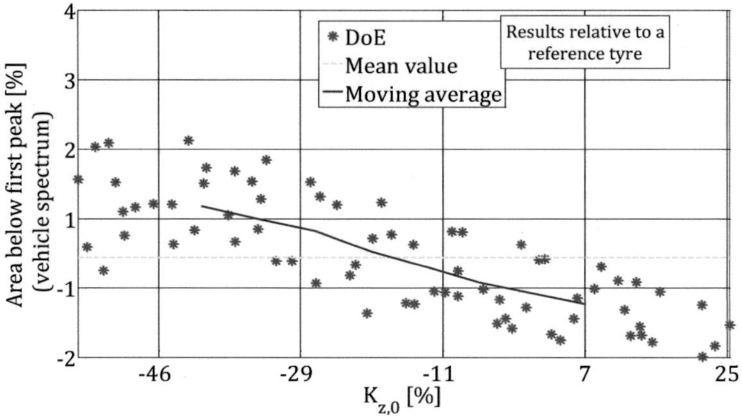

Figure 6.10: Graphical analysis, influence of dynamic vertical stiffness at zero hertz on the spectrum of the vehicle bouncing acceleration

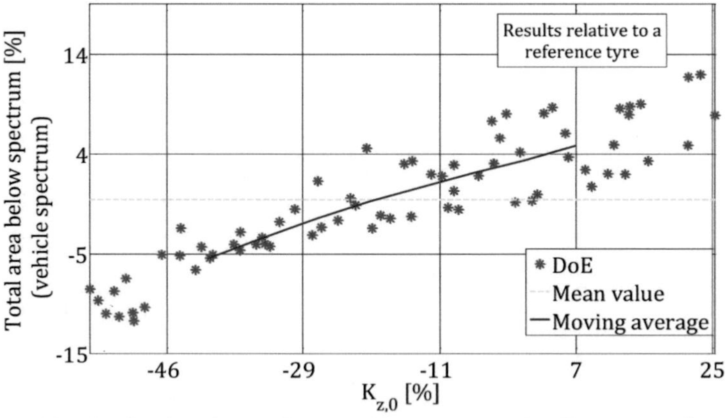

Figure 6.11: Graphical analysis, influence of dynamic vertical stiffness at zero hertz on the spectrum of the vehicle bouncing acceleration

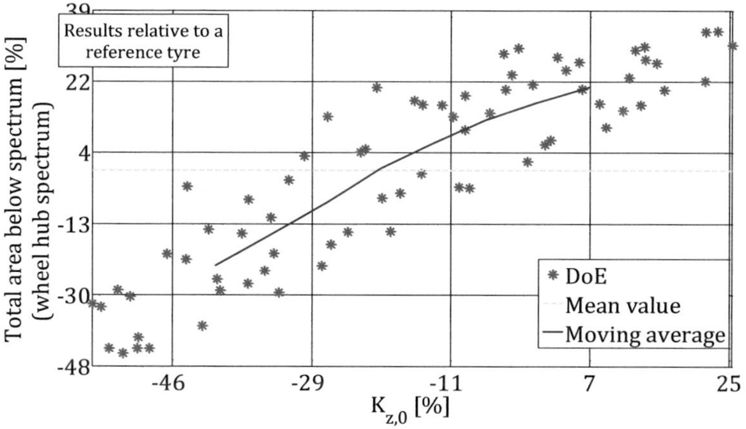

Figure 6.12: Graphical analysis, influence of dynamic vertical stiffness at zero hertz on the spectrum of the wheel hub acceleration

The relevant FTCs for ride comfort are enlisted in Table 6.13. Although the dynamic vertical stiffness $K_{z,0}$ dominates the influence on the ride comfort OMC, the slope $K'_{z,14,f}$ is added in order to describe better the whole characteristics of K_z between 0 to 30 Hz.

Table 6.13: Relevant FTCs for ride comfort

Name	Formulation	Symbol	Unit
Dynamic vertical stiffness at zero hertz	$K_z\vert_0$	$K_{z,0}$	$[N/m]$
Slope of dynamic vertical stiffness	$\left.\frac{\partial K_z}{\partial f}\right\vert_{14\,Hz}$	$K'_{z,14,f}$	$[Ns/m]$

6.3 Summary

In section 1 the requirement ride comfort is analysed. First, the choice of manoeuvres is motivated. The road sweep (see Table 6.1) allows to evaluate the spectrum of vehicle and wheel-suspension system in the frequency range up to 30 Hz.

Then, twelve relevant objective manoeuvre criteria (OMC) for ride comfort are presented (see Table 6.6): six describe the spectra of bouncing and pitching acceleration at the vehicle centre of mass; additional six describe the spectra of the vertical accelerations at the wheel hub of front and rear axle.

Finally, six functional tyre characteristics (FTCs) concerning ride comfort are introduced and a graphical sensitivity analysis is performed. The analysis is based on 75 virtual tyres generated by the Tyre Shaper: moreover, two different vehicles are taken into account. Two FTCs are chosen as relevant for describing ride comfort: the dynamic vertical stiffness $K_{z,0}$ and its slope $K'_{z,14,f}$ at 14 Hz (see Table 6.13).

7 Interior Noise

7.1 Objective Manoeuvre Criteria

In this section, first, the manoeuvres used to evaluate the interior noise of vehicles and the objective manoeuvre criteria (OMC) necessary to describe its characteristics are presented.

7.1.1 Choice of Manoeuvres

The manoeuvres for interior noise are defined to excite best the vehicle's eigenmodes: in fact, axles and vehicle body transfer the forces at the contact patch into the vehicle; their resonances amplify them generating higher interior noise. According to considerations of section 2.4.1 the frequency range for interior noise caused by the tyre-road interaction can be limited between 30 and 300 Hz. To this purpose the spectra of the vehicle's response are analysed.

Manoeuvres concerning interior noise are typically performed driving straight-forward under constant velocity or given acceleration (e.g. pass-by-noise under acceleration, coast down). In fact, the differences between the manoeuvres are given by the road excitation. Each road is characterized by a micro texture, macro texture and mega texture (Genuit, 2010). The first is too small to be observed by the eye and determines the grip of the surface; the second has wavelengths in the same order of magnitude as the tyre tread elements (see Figure 7.1); the third has wavelengths in the same order of magnitude as the tyre-road contact area and is materialised as potholes and "waviness". Bigger textures are referred to as "unevenness".

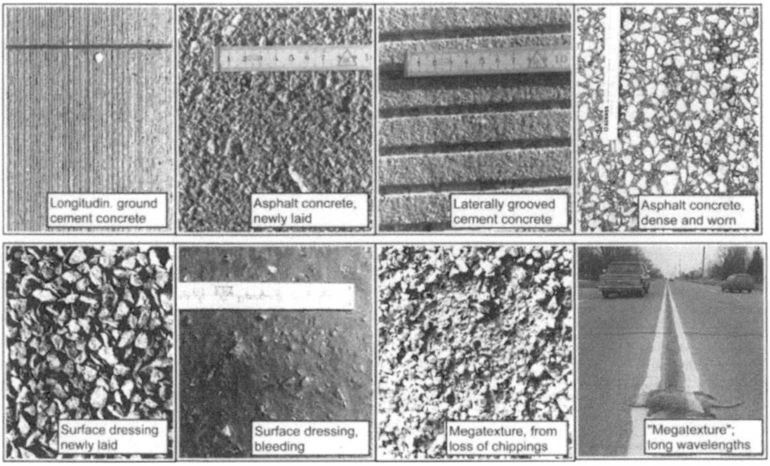

Figure 7.1: Examples of some macro textures (Sandberg & Ejsmont, 2002)

Two roads are chosen: the "rough road" and the "smooth road with unevenness". They excite both low-frequency and high-frequency acoustic phenomena. In order to isolate the contribution of the tyre-road interaction from the noise generated by the power unit and the wind turbulence, measurements are performed at low engine load (constant velocity on horizontal road), low engine speed (ca. 2000 rpm) and at low velocity (ca. 30 to 70 km/h).

7.1.1.1 Rough Road

The "rough road" (see Figure 7.2) is used to evaluate interior noise especially in the lower frequency range ($< 150\ Hz$). It is a 500 m long and 3 m wide road with a texture that generates high energy input at the contact patch: it excites the tyre at low frequencies, allowing to evaluate hum, low rumble and drone phenomena (see section 0). Manoeuvres on the rough road are performed driving straight-forward at speeds from 30 to 50 km/h.

Figure 7.2: Rough road

7.1.1.2 Smooth Road with Unevenness

The "smooth road with unevenness" (see Figure 7.3) is used to evaluate interior noise, especially in the higher frequency above 150 Hz. It is a 1000 m long and 4 m wide road with a smooth surface characterized by long wavelength unevenness that generate roll and bumping. Moreover, small impacts, e.g. gullies and bumps, excite the tyre at higher frequencies allowing to evaluate high rumble and impact damping phenomena (see section 0). Manoeuvres on the smooth road with unevenness are performed driving straight-forward at speeds from 50 to 70 km/h.

Figure 7.3: Smooth road with unevenness

7.1.2 Choice of Objective Manoeuvre Criteria

The OMC should describe the interior noise as heard from the driver and from a passenger sitting in the back of the vehicle. Interior vehicle noise perception is characterized by several different phenomena concerning sound pressure level (SPL), acoustic colour and tonal effects (see section 0).

Four objective manoeuvre criteria (OMC) are defined; they are enlisted in Table 7.4. The first two measure the average A-weighted SPL (see section 2.3.2) of the main interior noise phenomena: $\bar{S}_L$ for hum, low rumble and drone (30 to 100 Hz) and $\bar{S}_M$ for high rumble (100 to 200 Hz).

Table 7.4: Relevant OMC for interior noise

Characteristics	Formulation	Symbol	Description
Sound pressure level	$\frac{1}{70}\int_{30}^{100} \lvert p_A \rvert df$	$\bar{S}_L$	A-weighted bandpass for low frequency.
	$\frac{1}{100}\int_{100}^{200} \lvert p_A \rvert df$	$\bar{S}_M$	A-weighted bandpass for middle frequency.
Tonal effect	$\frac{1}{50}\int_{200}^{250} \lvert p_A \rvert df$	$\bar{S}_C$	A-weighted bandpass for air cavity noise.
Acoustic colour	$(\bar{S}_H + \bar{S}_C) - \bar{S}_L$	S_{col}	Relative SPL between low frequency and middle frequency noise.

The third OMC, $\bar{S}_C$, measures the A-weighted SPL at higher frequency: this frequency range is dominated by the tonal effect of the air-cavity resonance, occurring for most tyre dimensions between 200 and 250 Hz. The fourth characterizes the acoustic colour evaluating the contribution to the A-weighted SPL of low rumble in relation to the one of high rumble and air cavity. It should be noticed, that the A-weight is applied before evaluating the average value of the amplitudes of the sound pressure p_A in the selected bandpass range.

All OMC can be measured by microphones at the left or right ear of the driver or of a passenger sitting in the back of the vehicle: it is proven, that the OMC that correlate best with subjective evaluations are those measured at the driver's left ear. In Figure 7.5 the first three OMC are exemplarily displayed for a measurement of a middle-class coupé.

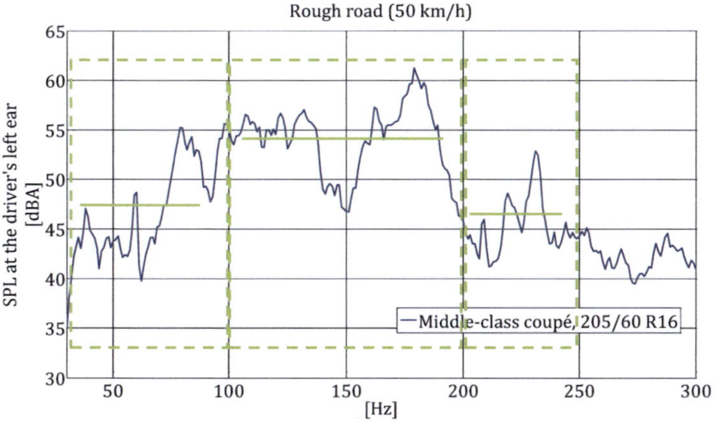

Figure 7.5: OMC for interior noise

7.2 Functional Tyre Characteristics

7.2.1 Definition of Functional Tyre Characteristics

The present thesis focuses on the interior noise generated by the tyre-road interaction. Subsequently, the FTCs presented in this section should allow an exhaustive description of the tyre force spectra at the wheel hub.

The spectra concern longitudinal, lateral and vertical forces. However, it should be noticed that bushings of wheel-suspension systems are softer in longitudinal direction than in radial direction (i.e. lateral and vertical direction). Moreover, longitudinal forces are in magnitude similar to vertical forces and on average two times bigger than lateral forces.

Seven FTCs are defined to describe the spectra: one for the longitudinal force, one for the lateral force and five for the vertical force (see Table 7.7). In Figure 7.6 the FTCs of the vertical force are shown.

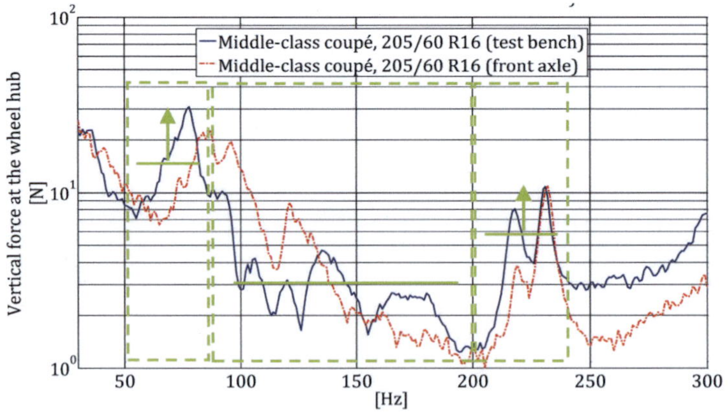

Figure 7.6: FTCs for vertical force

They are selected to describe best the force spectra at the wheel hub, especially the average force contributions of the eigenmodes and in the ranges between the eigenfrequencies. The FTC of the longitudinal force describes the average force between the frequencies $f_{x,1}$ and $f_{x,2}$ of respectively first and second peak at low frequency. The FTC of the lateral force focuses on the average force between the first eigenfrequency $f_{y,1}$ and 250 Hz. The vertical force is characterized by two eigenmodes: the first vertical tyre eigenmode (eigenfrequency $f_{z,1}$) and the one of the air cavity (eigenfrequency $f_{z,c}$). Subsequently, the ranges of FTCs describing their contributions are defined in accordance to the position of the eigenfrequency. Moreover, not only the average values are given, but also their relative force peaks ($\hat{F}_{z,V}$, $\hat{F}_{z,C}$).

Table 7.7: The FTCs for interior noise

Name	Formulation	Symbol	Unit
Average longitudinal force at low frequency	$\frac{1}{(f_{x,2}-20)-(f_{x,1}+20)}\int_{f_{x,1}+20}^{f_{x,2}-20}\lvert F_x\rvert df$	$\bar{F}_{x,L}$	[N]
Average lateral force at middle frequency	$\frac{1}{250-(f_{y,r1}+20)}\int_{f_{y,1}+20}^{250}\lvert F_y\rvert df$	$\bar{F}_{y,M}$	[N]
Average vertical force due to first eigenmode	$\frac{1}{40}\int_{f_{z,1}-20}^{f_{z,1}+20}\lvert F_z\rvert df$	$\bar{F}_{z,V}$	[N]
Average vertical force at middle frequency	$\frac{1}{(f_{z,c}-20)-(f_{z,1}+20)}\int_{f_{z,1}+20}^{f_{z,c}-20}\lvert F_z\rvert df$	$\bar{F}_{z,M}$	[N]
Average vertical force due to air cavity	$\frac{1}{40}\int_{f_{z,c}-20}^{f_{z,c}+20}\lvert F_z\rvert df$	$\bar{F}_{z,C}$	[N]
Relative force peak due to first vertical eigenmode	$\frac{\lvert F_z(f_{z,1})\rvert-\bar{F}_{z,V}}{\bar{F}_{z,V}}$	$\hat{F}_{z,V}$	[N]
Relative force peak due to air cavity	$\frac{\lvert F_z(f_{z,c})\rvert-\bar{F}_{z,C}}{\bar{F}_{z,C}}$	$\hat{F}_{z,C}$	[N]

7.2.2 Experimental Study for Interior Noise

The experimental study for interior noise focuses on the influence the vehicle and tyre architecture, the road texture and the vehicle operating conditions have on in-vehicle interior noise. The DoE is a full-factorial design (FFD) consisting of 96 on-track measurements and 48 test bench measurements (see Table 7.8).

The 205/60 R16 standard tyre is a rolling resistance optimised comfort tyre, while the 225/45 R18 run-flat tyre is developed for sport vehicles.

Table 7.8: Experimental study for interior noise

Parameter	Number	Comment
Tyres	2	205/60 R16 standard tyre (STD) 225/45 R18 run-flat tyre (RFT)
Vehicles	3	Compact vehicle, middle-class touring, middle-class coupé.
Test tracks	2	Rough road generating high forces at low frequencies. Smooth road with unevenness generating impacts and forces at higher frequencies.
Test bench	1	Outer drum test bench for acoustic measurements with rough surface.
Trajectory velocity	2	Rough road: 30 and 50 km/h. Smooth road with unevenness: 50 and 70 km/h.
Inflation pressure	2	Nominal inflation pressure for the selected tyre-vehicle combination and a 25 % higher inflation pressure.
Tyre load	2	Nominal tyre load given by the vehicle mass with two passengers and full tank and a 25 % higher tyre load.

The vehicles are chosen so, that all tyre-vehicle combinations are allowed for series production: they represent customer relevant combinations.

Two test tracks are chosen: the rough road (see section 7.1.1.1) and the smooth road with unevenness (see section 7.1.1.2). Measurements are performed at different velocities, inflation pressures and tyre loads (tyre load is increased with sandbags).

The measurement equipment consists of four microphones (see Figure 7.9) and four dynamometric wheels (see Figure 7.10). The microphones are positioned near the driver's left and right ear, as well as near the left and right ear of the passenger sitting in the back. The analyses focus on the SPL measured at the driver's left ear. The dynamometric wheels measure forces and moments at each wheel hub: their weight is similar to that of a standard aluminium wheel ($+4\ kg$). They allow high resolution measurements in the frequency range of interest (till 350 Hz). For all six tyre-vehicle combinations also test bench measurements are performed at different velocities and inflation pressures. The tyre load is chosen as average tyre load of front and rear axle of each the vehicle. The acoustic test bench in shown in Figure 7.11.

Figure 7.9: Microphones for driver and passenger in the back

Figure 7.10: Dynamometric wheel

During a standard tyre development, tyres are evaluated at the test bench and not on a test track with dynamometric wheels. Therefore, a comparison between test bench and test track is performed. The results are shown in Figure 7.12.

Figure 7.11: BMW acoustic tyre test bench

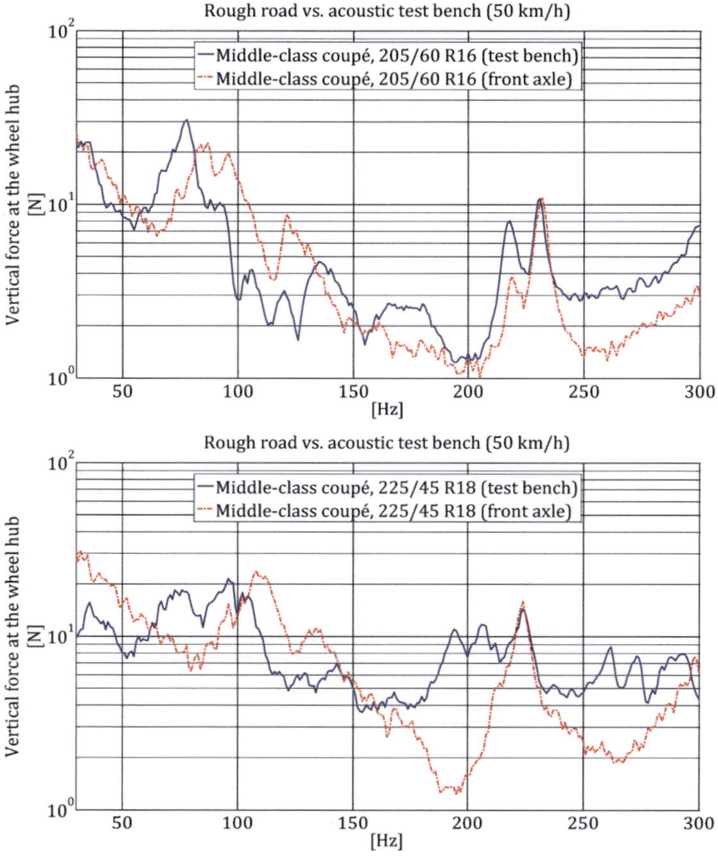

Figure 7.12: Rough road vs. acoustic test bench measurements

The measurements are fairly comparable: the eigenfrequency of the resonances are different, but shape and peak values are similar, especially the one of the cavity mode. The differences are given mainly by the absence of the wheel-suspension system and the different textures between road and test bench. Concluding, measurements on the test bench can be used to estimate the amplitudes of on-track measured force spectra.

7.2.3 Influence of Operating Conditions on the SPL

Before correlating the FTCs to SPL the influence of the operating conditions on the SPL is analysed, among others road texture and inflation pressure. The results described in this section are developed in collaboration with (Sochor, 2014): however, instead of focusing only on one vehicle, the conclusions taken in this section rely on measurements concerning three different vehicles.

The vehicle body as well as the axle architecture have significant influence on the SPL: average level as well as shape of the spectra change (see Figure 7.13). The middle-class coupé has high SPL values in the middle frequency range (100 to 200 Hz), while the compact vehicle is particularly sensible to the air cavity resonance and the middle-class touring to low frequencies (30 to 100 Hz).

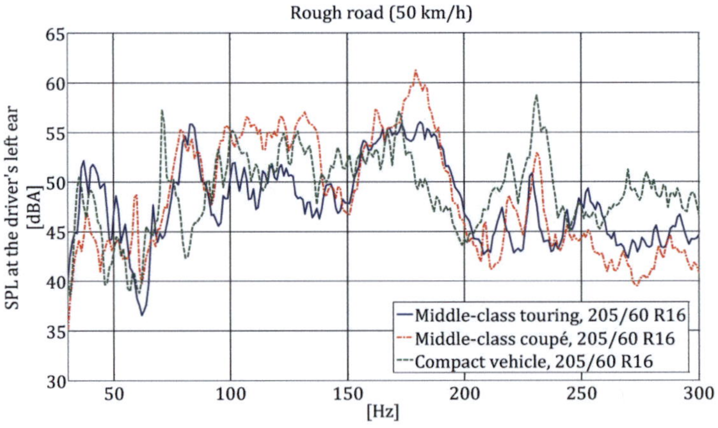

Figure 7.13: Influence of vehicle on SPL

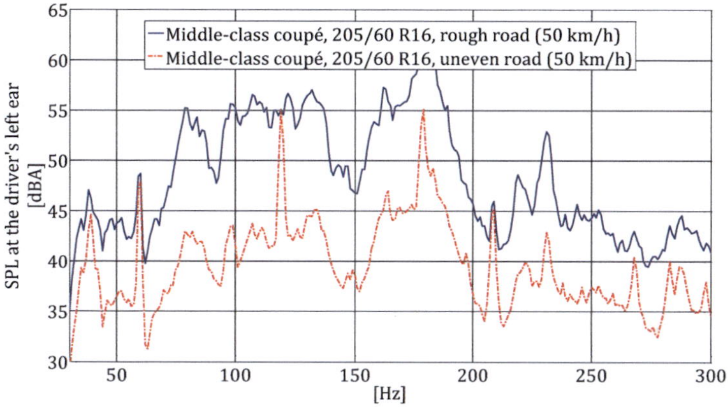

Figure 7.14: Influence of road texture on SPL

Road texture has a big influence on the SPL as well as on the shape of the spectra (see Figure 7.14). As mentioned before, the rough road generates high energy input, especially at low frequencies, while the smooth road with unevenness has a smooth surface exciting the tyre at higher frequencies with small impacts, e.g. gullies and bumps.

Interior noise is generated by two contributions: structure-borne noise, influenced by the tyre eigenmodes, and air-borne noise, influenced by air-pumping and aerodynamic phenomena (see section 2.4.1). It is known, that air-borne noise gains importance with frequency. In fact, it could be noticed, that above 250 *Hz* the vehicle architecture dominates the SPL of interior noise: independently from the operating conditions and from tyre type, it is possible to cluster the interior noise of the three vehicles (see Figure 7.15).

Higher trajectory velocities increase the excitation frequency of the road and the dynamic forces at the wheel, causing the average SPL to rise. In Figure 7.16 the effect of trajectory velocity is exemplarily shown for the 205/60 R16 standard tyre.

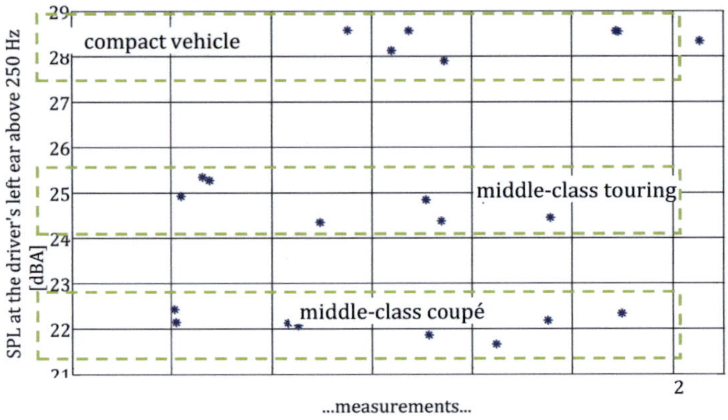

Figure 7.15: Acoustic cluster of vehicles

Higher inflation pressures increase the stiffness of the tyre, causing the peaks of the SPL to shift to higher frequencies (especially the peak around 80 Hz) and the average SPL to rise slightly. In Figure 7.17 the influence of inflation pressure is exemplarily shown for the 205/60 R16 standard tyre.

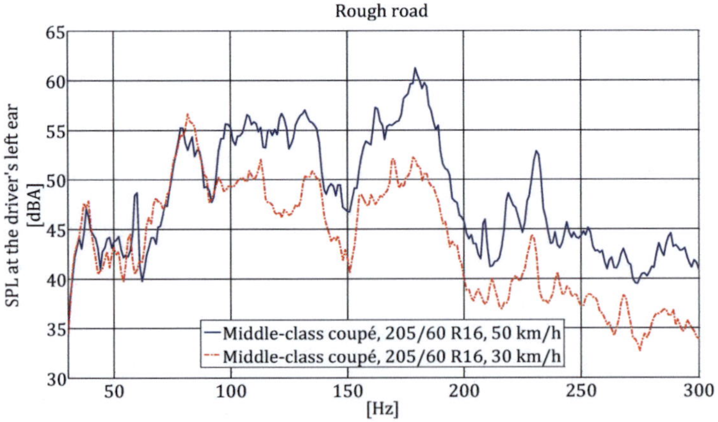

Figure 7.16: Influence of trajectory velocity on SPL

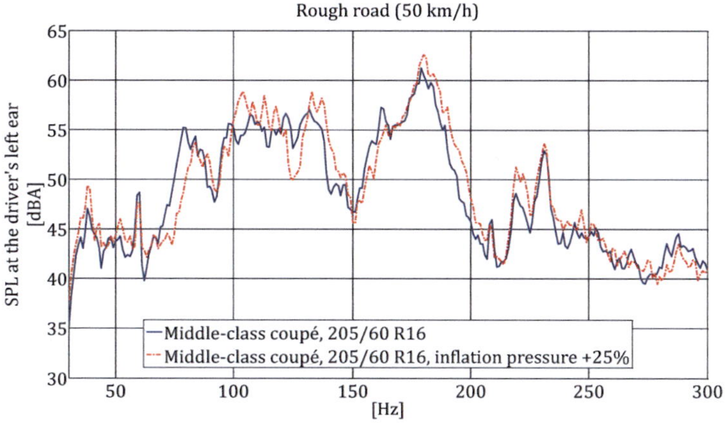

Figure 7.17: Influence of inflation pressure on SPL

Concerning tyre load, no significant influence is observed. In Figure 7.18 the effect of tyre load is exemplarily shown for the 205/60 R16 standard tyre.

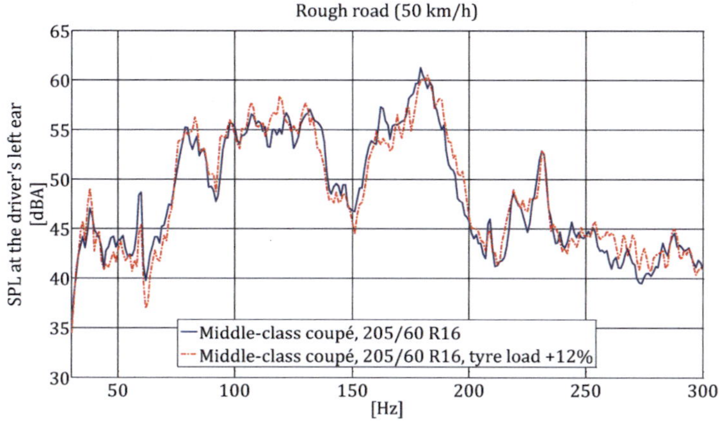

Figure 7.18: Influence of tyre load on SPL

7.2.4 Choice of Functional Tyre Characteristics

The choice of the relevant FTCs is done according to a graphical sensitivity analysis (see section 3.5.2.2) based on a design of experiment (DoE) consisting of 96 measurements.

The experiments of the DoE, a full-factorial design (FFD), are measurements of different tyre-vehicle combinations and operating conditions (see Table 7.8). The design variables of the DoE are the seven FTCs; the design targets are the four OMC as measured by microphones at the driver's left ear. The most significant results of the graphical analyses are presented in this section: the correlations are shown between FTCs measured at the test bench and OMC measured on the test track, knowing that this approach decreases the correlation value. However, this is allowed as FTCs from dynamometric wheel measurements correlate well with FTCs at the test bench (see Figure 7.12).

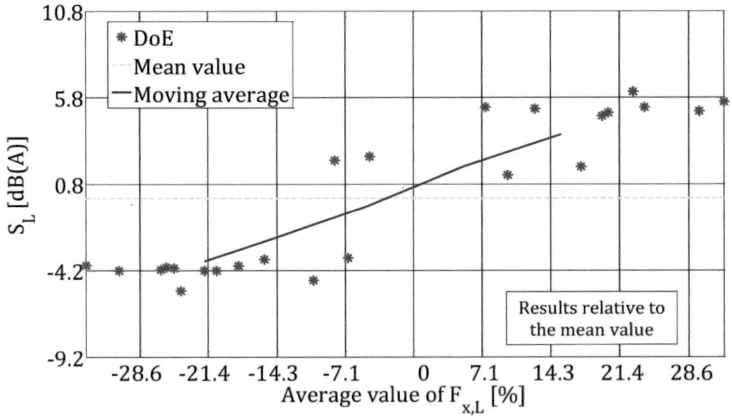

Figure 7.19: Graphical analysis for interior noise, influence of longitudinal force amplitudes on the SPL at low frequencies $\bar{S}_L$

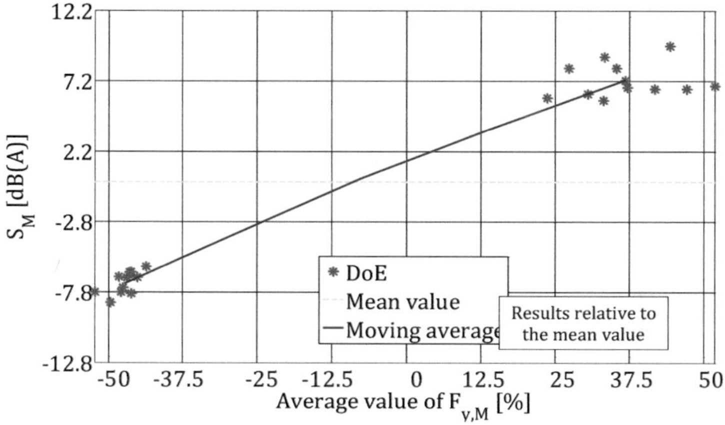

Figure 7.20: Graphical analysis for interior noise, influence of lateral force amplitudes on the SPL at middle frequencies $\bar{S}_M$

Concerning the average longitudinal force amplitudes, $\bar{F}_{x,L}$ correlates best with the normalised band-pass $\bar{S}_L$ between 30 and 100 Hz (see Figure 7.19).

Concerning the average lateral force amplitudes, $\bar{F}_{y,M}$ correlates best with the normalised band-pass $\bar{S}_M$ between 100 and 200 Hz (see Figure 7.20). It should be noticed, that the worst measurements (upper right corner of the figure) are performed with the 225/45 R18 run-flat tyre, which exhibits its particularly high lateral force amplitudes.

Concerning the vertical force, the average vertical force amplitudes $\bar{F}_{z,V}$ does not correlate with any OMC, while $\bar{F}_{z,M}$ and $\bar{F}_{z,C}$ correlate best with respectively the normalised band-pass $\bar{S}_M$ between 100 and 200 Hz and normalised band-pass $\bar{S}_C$ between 200 and 250 Hz (see Figure 7.21 and Figure 7.22). Moreover, both $\bar{F}_{z,M}$ and $\bar{F}_{z,C}$ influence the acoustic colour as expressed by S_{col}. Again, the 225/45 R18 tyre exhibits the worst results in terms of interior noise.

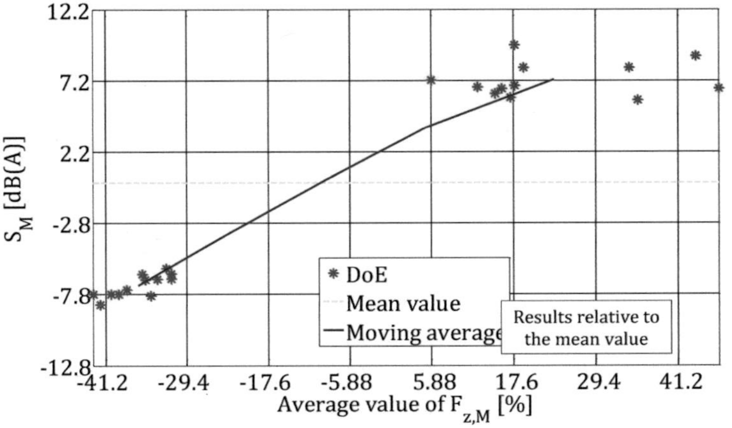

Figure 7.21: Graphical analysis for interior noise, influence of vertical force amplitudes on the SPL at middle frequencies $\bar{S}_M$

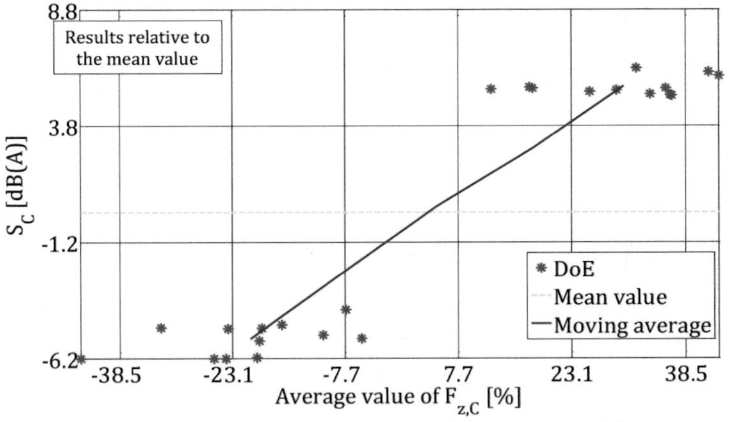

Figure 7.22: Graphical analysis for interior noise, influence of vertical force amplitudes on the SPL due to air cavity resonance $\bar{S}_C$

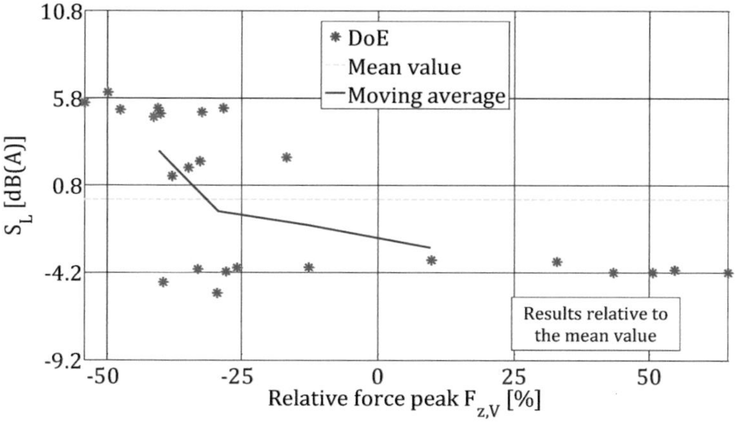

Figure 7.23: Graphical analysis for interior noise, influence relative force peak due to the first vertical eigenmode on the SPL at low frequencies $\bar{S}_M$

Concerning the relative force peak due to the first vertical eigenmode $\hat{F}_{z,M}$, it should be noted that high peak values generate less interior noise $\bar{S}_L$ (see Figure 7.23). This is of course not true: the FTC $\hat{F}_{z,M}$ is defined as ratio of peak amplitude to average amplitudes and, de facto, the low average force amplitudes are responsible for the relation and not the peak value. Similar considerations are valid for the relative force peak due to air cavity $\hat{F}_{z,C}$. Subsequently, the introduction of a FTC describing the relative force peak is not meaningful.

The relevant FTCs for interior noise are enlisted in Table 7.24. According to the results of the graphical analysis, the dominant FTCs are $\bar{F}_{x,L}$ and $\bar{F}_{y,M}$s well as $\bar{F}_{z,C}$ and $\bar{F}_{z,M}$. The average vertical force due to the first eigenmode $\bar{F}_{z,V}$ is added in order to describe better the spectrum of the vertical force. Moreover, it should be remembered, that the choice is valid for different vehicles: a compact vehicle, middle-class touring and a middle-class coupé.

Table 7.24: Relevant FTCs for interior noise (defined at a given tyre load $F_{z,ref}$ and inflation pressure p)

Name	Formulation	Symbol	Unit
Average longitudinal force at low frequencies	$\frac{1}{(f_{x,2}-20)-(f_{x,1}+20)}\int_{f_{x,1}+20}^{f_{x,2}-20}\lvert F_x\rvert df$	$\bar{F}_{x,L}$	$[N]$
Average lateral force at middle frequencies	$\frac{1}{250-(f_{y,1}+20)}\int_{f_{y,1}+20}^{250}\lvert F_y\rvert df$	$\bar{F}_{y,M}$	$[N]$
Average vertical force due to first eigenmode	$\frac{1}{40}\int_{f_{z,1}-20}^{f_{z,1}+20}\lvert F_z\rvert df$	$\bar{F}_{z,V}$	$[N]$
Average vertical force at middle frequencies	$\frac{1}{(f_{z,c}-20)-(f_{z,1}+20)}\int_{f_{z,1}+20}^{f_{z,c}-20}\lvert F_z\rvert df$	$\bar{F}_{z,M}$	$[N]$
Average vertical force due to air cavity	$\frac{1}{40}\int_{f_{z,c}-20}^{f_{z,c}+20}\lvert F_z\rvert df$	$\bar{F}_{z,C}$	$[N]$

7.3 Summary

In section 1 the requirement interior noise is analysed. First, the choice of manoeuvres is motivated. The "rough road" and the "smooth road with unevenness" (see section 7.1.1) allow to excite the vehicle's eigenmodes and to evaluate the interior noise in the frequency range between 30 and 300 Hz.

Then, four relevant objective manoeuvre criteria (OMC) for interior noise are presented (see Table 7.4): one describes the sound pressure level (SPL) of low frequency (hum, low rumble, drone) and two the SPL of middle-to-high frequency phenomena (high rumble, air cavity) and one the acoustic colour, i.e. the ratio between low frequency and high frequency interior noise.

Finally, seven functional tyre characteristics (FTCs) concerning interior noise are introduced and a graphical sensitivity analysis is performed. The analysis is based on 96 measurements at the acoustic tyre test bench and on a test track: different tyre-vehicle combinations as well as road textures and operating conditions are taken into account. Five FTCs are chosen as relevant for influencing interior noise (see Table 7.24).

8 Physical Tyre Model

In order to analyse the conflicts between requirements concerning power loss, lateral dynamics, comfort and interior noise, a tyre model, which is based on parameters common to all four requirements, is needed. Therefore, a physical tyre model representing the FTCs as a function of the tyre design parameters (TDPs) is developed: the tyre model is presented in this section.

8.1 Model Description

The physical tyre model is composed of three main sub-models (see Figure 8.9): a three mass model used to describe ride comfort and interior noise (in vertical and lateral direction), a brush model used to describe the lateral dynamics and a hysteresis model used to describe the rolling resistance.

In this section, first, the TDPs are introduced and the geometrical and material properties are described. Then, the equations yielding the static equilibrium are derived. Finally, the equation of motions for ride comfort, interior noise, lateral dynamics and rolling resistance are presented.

8.1.1 Tyre Design Parameter

A list of the TDPs is given in Table 8.1: the enlisted parameters are independent and allow a unique description of the tyre dynamics (rolling resistance, lateral dynamics, ride comfort and interior noise). The values of the parameter are an example and refer to a 205/55 R16 run-flat tyre (Continental, 2003).

The geometry of the tyre is described by seven parameters (from R to e_r). Due to complexity, the geometry of the tread design is disregarded: only information about the void ratio is given. Subsequently, it is necessary to adjust the elastic modulus of the tread in order to include the contribution of the tread design to the tread's stiffness in lateral and vertical direction. The material properties of the tyre are described by twelve parameters (from E_s to m). All stiffnesses concerning the tyre are a function of the geometrical properties and the elastic moduli. The operating conditions of the tyre are described by four parameters (from p to $\mu_{a,ref}$). They are not a property of the tyre and can be influenced by the customer (e.g. vehicle load) and by the environment (e.g. wet road). The six form factors allow to take into account the influence of components that are disregarded as of minor importance (e.g. inner liner, filler, chafer, ...), as well as of the construction (e.g. tyre contour) and of the production process of the tyre. They affect vertical stiffness k_s of sidewall and k_p due to inflation pressure (f_{k_s}, f_{k_p}), shear modulus G_{ys} of the sidewalls $(f_{G_{ys}})$, total shear modulus G_y of the tyre (f_{G_y}) and damping coefficients d_{y1}, d_{y2}, d_{z1} and d_{z2} of sidewall, belt and tread (f_{d_y}, f_{d_z}).

Table 8.1: TDPs (exemplarily for 205/55 R16 run-flat tyre)

Name	Symbol	Unit	Value
Tyre radius	R	$[m]$	$310/1000$
Nominal width	w_n	$[m]$	$205/1000$
Sidewall height	h_s	$[m]$	$110/1000$
Sidewall thickness	w_s	$[m]$	$15/1000$
Belt height	h_b	$[m]$	$7/1000$
Tread height	h_t	$[m]$	$7/1000$
Void ratio	e_r	$[-]$	$35/100$
Elastic modulus of sidewall	E_s	$[N/m^2]$	$3.0 \cdot 10^6$
Elastic modulus of belt	E_b	$[N/m^2]$	$2.0 \cdot 10^9$
Elastic modulus of tread	E_t	$[N/m^2]$	$2.0 \cdot 10^6$
Rim stiffness	k_r	$[N/m]$	$30 \cdot 10^6$
Air stiffness	k_a	$[N/m]$	$70 \cdot 10^3$

Name	Symbol	Unit	Value
Rim damping	d_r	$[Ns/m]$	15
Air damping	d_a	$[Ns/m]$	0.10
Sidewall damping	d_s	$[Ns/m]$	35
Belt damping	d_b	$[Ns/m]$	20
Tread damping	d_t	$[Ns/m]$	30
Rim mass	m_r	$[kg]$	9
Tyre mass	m	$[kg]$	10
Inflation pressure	p	$[Pa]$	$2.2 \cdot 10^5$
Tyre load	F_z	$[N]$	$4.0 \cdot 10^3$
Trajectory velocity	v_x	$[m/s]$	80 / 3.6
Reference adhesion coefficient	$\mu_{a,ref}$	$[-]$	2.70
Form factor 1	f_{k_s}	$[-]$	0.9
Form factor 2	f_{k_p}	$[-]$	0.5
Form factor 3	$f_{G_{ys}}$	$[-]$	40
Form factor 4	f_{G_y}	$[-]$	10
Form factor 5	f_{d_y}	$[-]$	1.5
Form factor 6	f_{d_z}	$[-]$	10
Reference ground pressure	p_{ref}	$[Pa]$	$2.5 \cdot 10^5$
Exponent for pressure distribution	q_1	$[-]$	2.5
Exponent for friction coefficient distribution	q_2	$[-]$	0.3
Ratio between reference adhesion and sliding coefficient	q_3	$[-]$	0.55

8.1.2 Geometrical Properties

In this section, the equations describing the geometrical properties as a function of the TDPs are presented.

8.1.2.1 Width and Radii

Belt width w_b and tread width w_t are approximately 15 % smaller than the nominal width w_n (Continental, 2003):

$$w_b = 0.85 \cdot w_n \tag{8.1}$$

$$w_t = 0.85 \cdot w_n \tag{8.2}$$

Rim radius R_r is derived according to geometrical considerations:

$$R_r = R - h_s - h_b - h_t \tag{8.3}$$

In order to model the tyre curvature in lateral direction two radii are introduced: $R_{y,t}$ for the tread and $R_{y,b}$ for the belt. Figure 8.2 shows the lateral section of the tyre, the radii $R_{y,i}$, the radial deflections δ_i and the widths w_i.

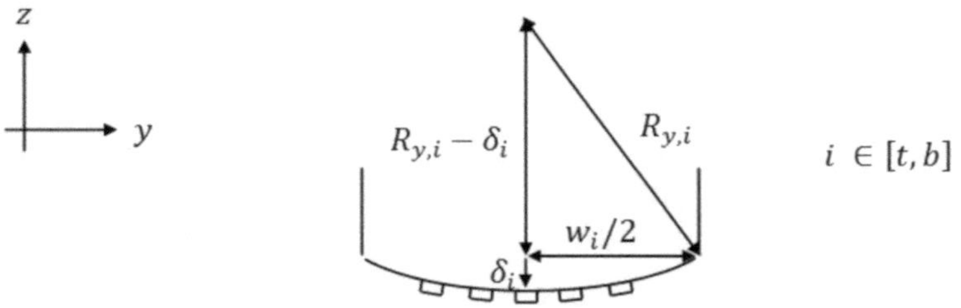

Figure 8.2: Tyre curvature in lateral direction

The radius $R_{y,i}$ describing the curvature of the tread (index t) or belt (index b) is defined as:

$$R_{y,i} = \frac{\left(\frac{w_i}{2}\right)^2 + \delta_i^2}{2\delta_i} \qquad i \in [t, b] \tag{8.4}$$

where δ_i is the deflection of the tyre in radial direction. For inflation pressures around nominal conditions ($p \in [1.5, 3.5] \cdot 10^5\ Pa$) deflection δ_i is assumed to be related to inflation pressure according to the following empirical formulation:

$$\delta_i = \delta_{i,n} \left(\frac{p}{p_n}\right)^{0.8} \qquad i \in [t, b] \qquad \textbf{8.5}$$

where $\delta_{i,n}$ is tread deflection at the nominal inflation pressure p_n. Reasonable deflections for a 205/55 R16 run-flat tyre at a nominal inflation pressure of $2.2 \cdot 10^5\ Pa$ are $\delta_{t,n} = 9.0\ mm$ and $\delta_{b,n} = 6.5\ mm$.

8.1.2.2 Deflection of the Tyre

Figure 8.3 and Figure 8.4 show respectively a non-deformed tyre and a deformed tyre. z_h, z_r and z_b are independent variables describing the vertical displacements of respectively wheel hub, rim and belt. z describes the vertical displacement of the road surface.

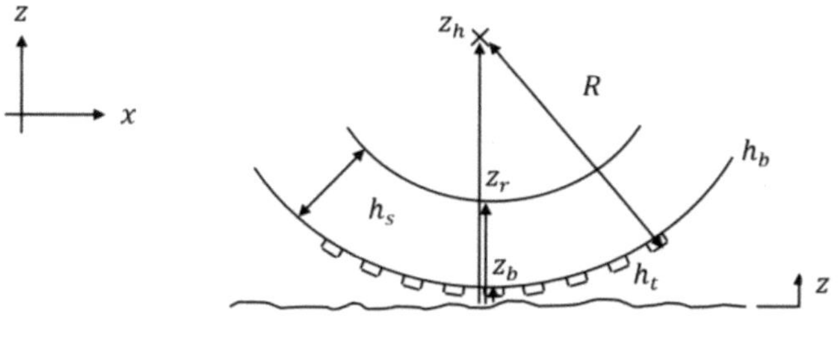

Figure 8.3: Non-deformed tyre

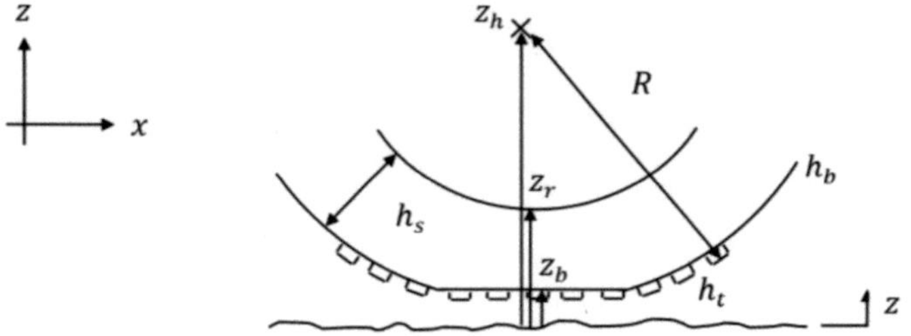

Figure 8.4: Deformed tyre

The contact area $A_{c,xy}$ between tread and road is non-zero for the deformed tyre. Contact area $A_{c,xy}$ differs from pressure area $A_{p,xy}$ (the area on which inflation pressure works) due to the presence of the belt and the tread. In Figure 8.5 the difference between the lengths of the two areas is shown.

Concerning the contact area, width w_c is limited between zero and nominal width w_n:

$$w_c = min\left(2\sqrt{2R_{y,t}z - z^2}, w_n\right) \qquad z \geq 0 \qquad \mathbf{8.6}$$

$$w_c = 0 \qquad z < 0 \qquad \mathbf{8.7}$$

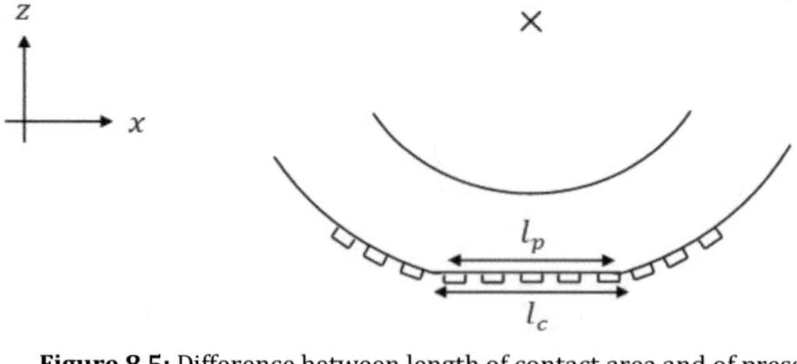

Figure 8.5: Difference between length of contact area and of pressure area

Similarly, length l_c is:

$$l_c = 2\sqrt{2Rz - z^2} \qquad z \geq 0 \qquad \textbf{8.8}$$

$$l_c = 0 \qquad z < 0 \qquad \textbf{8.9}$$

Concerning the pressure area, width w_p is limited between zero and nominal width w_n:

$$w_p = min\left(2\sqrt{2R_{y,b}z_b - {z_b}^2}, w_n\right) \qquad z_m \geq 0 \qquad \textbf{8.10}$$

$$w_p = 0 \qquad z_m < 0 \qquad \textbf{8.11}$$

Length l_p is:

$$l_p = 2\sqrt{2(R - h_b - h_t)z_b - z_b^2} \qquad z_m \geq 0 \qquad \textbf{8.12}$$

$$l_p = 0 \qquad z_m < 0 \qquad \textbf{8.13}$$

8.1.2.3 Areas

Contact area $A_{c,xy}$ is defined as:

$$A_{c,xy} = \frac{\pi}{4} \cdot w_c l_c \cdot (1 - e_r) \qquad \textbf{8.14}$$

where $\pi/4$ is a factor that takes into account the elliptical form of the contact area ($\pi/4$ is the ratio of the areas of a rectangle and an ellipse) and e_r is the void ratio. Similarly, pressure area $A_{p,xy}$ is defined as:

$$A_{p,xy} = \frac{\pi}{4} \cdot w_p l_p \qquad \textbf{8.15}$$

Finally, the areas of sidewall sections $A_{s,xy}$ and $A_{s,xz}$ are defined as:

$$A_{s,xy} = w_s l_p \qquad \textbf{8.16}$$

$$A_{s,xz} = h_s l_p \quad 8.17$$

In (Fujikawa, Funazaki, & Yamazaki, 1994) it is shown, that due to road roughness the effective contact area is smaller than the nominal contact area (here approximated by $A_{c,xy}$). Due to complexity and a negligible improvement in accuracy no correction coefficient is introduced.

8.1.2.4 Geometrical Moments of Inertia

The moments of inertia of sidewall, belt and tread are approximated by rotations of rectangular cross sections around their centred axis:

$$I_{s,yx} = \frac{w_s^3 l_p}{12} \quad 8.18$$

$$I_{s,zy} = \frac{h_s^3 w_s}{12} \quad 8.19$$

$$I_{b,zy} = \frac{h_b^3 w_b}{12} \quad 8.20$$

$$I_{t,zy} = \frac{h_t^3 w_t}{12} \quad 8.21$$

8.1.3 Material Properties

In this section, the equations describing the material properties as a function of the TDPs are presented. For deriving the relation between vertical stiffness, elastic modulus and shear modulus, the assumption is made, that sidewall, belt and tread are isotropic, linear-elastic and non-auxetic materials.

8.1.3.1 Total Vertical Stiffness

Total vertical stiffness k is the sum of three stiffnesses in series:

$$k(z, z_b, z_r) = \left({k_r}^{-1} + {k_{z1}}^{-1} + {k_{z2}}^{-1}\right)^{-1} \qquad \textbf{8.22}$$

where k_r is the rim stiffness, k_{z1} the stiffness of sidewalls and due to inflation pressure and k_{z2} the stiffness of belt and tread.

8.1.3.2 Vertical Stiffness of the Rim

Rim stiffness k_r (see Figure 8.9) is assumed to be constant with frequency and amplitude (small displacements):

$$k_r = \frac{\partial F_r}{\partial(z_h - z_r)} \qquad \textbf{8.23}$$

where F_r is the force acting at the rim hub (at static equilibrium equal to the tyre load F_z). The force is negative, if the rim is compressed.

8.1.3.3 Vertical Stiffness of Sidewalls and Inflation Pressure

Vertical stiffness k_{z1} (see Figure 8.9) of sidewalls and inflation pressure is assumed to be constant with frequency and amplitude (small displacements):

$$k_{z1} = \frac{\partial F_{z1}}{\partial(z_r - z_b)} \qquad \textbf{8.24}$$

where F_{z1} is the vertical force acting between belt and rim (at static equilibrium equal to the tyre load F_z). It is the sum of three stiffnesses in parallel:

$$k_{z1} = 2 \cdot k_s + k_p \qquad \textbf{8.25}$$

Sidewall stiffness is assumed to be linear with deflection (isotropic, linear-elastic and non-auxetic material):

$$k_s = \frac{\partial F_s}{\partial(z_r - z_b)} = f_{k_s} \cdot \frac{E_s A_{s,xy}}{h_s} \qquad \textbf{8.26}$$

where F_s is the force generated by the sidewall and E_s, $A_{s,xy}$ and h_s are respectively the elastic modulus, the area and the height of the sidewall.

The form factor f_{k_s} is introduced in order to take into account the reduction of vertical stiffness due to lateral bending of the sidewall. The stiffness generated by the inflation pressure is:

$$k_p = \frac{\partial F_p}{\partial(z_r - z_b)} = f_{k_p} \cdot p \frac{\partial A_{p,xy}}{\partial(z_r - z_b)} \qquad \textbf{8.27}$$

where F_p is the force generated by inflation pressure (assumed constant for small displacements) and f_{k_p} is a form factor taking into account the circular architecture of the tyre.

8.1.3.4 Vertical Stiffness of Belt and Tread

Vertical stiffness k_{z2} (see Figure 8.9) of belt and tread is assumed to be constant with frequency and amplitude (small displacements):

$$k_{z2} = \frac{\partial F_{z2}}{\partial(z_b - z)} \qquad \textbf{8.28}$$

where F_{z2} is the vertical force acting between road and belt (at static equilibrium equal to the tyre load F_z). Vertical stiffness k_{z2} is expressed as the sum of two stiffnesses in series:

$$k_{z2} = \left(k_b{}^{-1} + k_t{}^{-1}\right)^{-1} \qquad \textbf{8.29}$$

The vertical stiffness of the belt and the tread are assumed to be linear with deflection (isotropic, linear-elastic and non-auxetic material):

$$k_b = \frac{\partial F_b}{\partial(z_b - z)} = \frac{E_b A_{c,xy}}{h_b} \qquad \textbf{8.30}$$

$$k_t = \frac{\partial F_t}{\partial(z_b - z)} = \frac{E_t A_{c,xy}}{h_t} \qquad \textbf{8.31}$$

where F_b and F_t are the forces acting on belt and tread, E_b and E_t the elastic moduli and h_b and h_t the height of belt and tread.

8.1.3.5 Shear Modulus

For an isotropic, linear-elastic and non-auxetic material the vertical stiffness can be related to the elastic modulus:

$$k_z = \frac{EA}{h} \quad \mathbf{8.32}$$

and the shear modulus to the elastic modulus:

$$G = \frac{\tau}{\gamma} = \frac{1}{2(1+\nu)}E \quad \mathbf{8.33}$$

where τ is the shear stress, γ the shear strain and ν Poisson's ratio non-negative and not bigger than 0.5, so that:

$$\frac{1}{3}E < G < \frac{1}{2}E \quad \mathbf{8.34}$$

Subsequently, the lateral stiffness is related to the shear modulus

$$k_y = \frac{GA}{h} \quad \mathbf{8.35}$$

Starting from the definition of stiffnesses in parallel

$$k^* = \sum_i^n k_i \quad \mathbf{8.36}$$

it is possible to derive an expression for the sum of n shear moduli "in parallel":

$$G^* = \frac{h^*}{A^*} \cdot \frac{1}{n} \sum_i^n \frac{G_i A_i}{h_i} \quad \mathbf{8.37}$$

where height h^* and area A^* are weighted average values:

$$h^* = \left(\sum_i^n \frac{G_i}{h_i}\right)^{-1} \cdot \sum_i^n G_i \quad \mathbf{8.38}$$

$$A^* = \sum_i^n (G_i A_i) \cdot \left(\sum_i^n G_i\right)^{-1} \qquad \textbf{8.39}$$

Similarly, the sum of n shear moduli "in series" is:

$$G^* = \frac{h^*}{A^*}\left(\frac{1}{n}\sum_i^n \left(\frac{G_i A_i}{h_i}\right)^{-1}\right)^{-1} \qquad \textbf{8.40}$$

where height h^* and area A^* are weighted average values:

$$h^* = \sum_i^n \left(\frac{G_i}{h_i}\right)^{-1} \cdot \left(\sum_i^n G_i^{-1}\right)^{-1} \qquad \textbf{8.41}$$

$$A^* = \left(\sum_i^n (G_i A_i)^{-1}\right)^{-1} \cdot \sum_i^n (G_i^{-1}) \qquad \textbf{8.42}$$

8.1.3.6 Total Shear Modulus

The total shear modulus G_y of the tyre is the sum of two shear moduli "in series":

$$G_y = f_{G_y} \cdot \frac{h^*_{Gy}}{A^*_{Gy}} \cdot \left(\frac{1}{2}\left(\left(\frac{G_{y1} A^*_{Gy1}}{h^*_{Gy1}}\right)^{-1} + \left(\frac{G_{y2} A^*_{Gy2}}{h^*_{Gy2}}\right)^{-1}\right)\right)^{-1} \qquad \textbf{8.43}$$

where f_{G_y} is a form factor taking into account the tyre structure, G_{y1} the shear modulus of sidewall and due to inflation pressure, and G_{y2} the shear modulus of belt and tread. Height h^*_{Gy} and area A^*_{Gy} are assumed to be (see section 8.1.3.5):

$$h^*_{Gy} = \left(\left(\frac{G_{y1}}{h^*_{Gy1}}\right)^{-1} + \left(\frac{G_{y2}}{h^*_{Gy2}}\right)^{-1}\right) \cdot \left(G_{y1}^{-1} + G_{y2}^{-1}\right)^{-1} \qquad \textbf{8.44}$$

$$A^*_{Gy} = \left(\left(G_{y1} A^*_{Gy1}\right)^{-1} + \left(G_{y2} A^*_{Gy2}\right)^{-1}\right)^{-1} \cdot \left(G_{y1}^{-1} + G_{y2}^{-1}\right) \qquad \textbf{8.45}$$

8.1.3.7 Shear Modulus of Sidewalls and Inflation Pressure

The shear modulus G_{y1} of the two sidewalls and due to inflation pressure is the sum of three shear moduli "in parallel":

$$G_{y1} = \frac{h^*_{Gy1}}{A^*_{Gy1}} \cdot \frac{1}{3}\left(2\frac{G_{ys}A_{s,xy}}{h_s} + \frac{G_{yp}A_{p,xy}}{h_s}\right) \qquad \textbf{8.46}$$

where G_{ys} is the shear modulus of the sidewall, G_{yp} the shear modulus due to inflation pressure. Height h^*_{Gy1} and area A^*_{Gy1} are assumed to be (see section 8.1.3.5):

$$h^*_{Gy1} = \left(2\frac{G_{ys}}{h_s} + \frac{G_{yp}}{h_s}\right)^{-1} \cdot \left(2G_{ys} + G_{yp}\right) \qquad \textbf{8.47}$$

$$A^*_{Gy1} = \left(2G_{ys}A_{s,xy} + G_{yp}A_{p,xy}\right) \cdot \left(2G_{ys} + G_{yp}\right)^{-1} \qquad \textbf{8.48}$$

The magnitude of the shear modulus due to inflation pressure can be derived approximating the tyre with a parallelogram (see Figure 8.6). The distributed force due to inflation pressure are represented by concentrated forces on tyre belt $F_{p,b}$ and tyre sidewalls $F_{p,s1}$, respectively acting in the middle of the belt width and the sidewall height:

$$\begin{aligned} F_{p,b} &= p \cdot A_{p,xy} \\ F_{p,s} &= p \cdot A_{s,xz} \end{aligned} \qquad \textbf{8.49}$$

Solving the structure the forces acting at the bounds are:

$$\begin{aligned} F_{A,y} &= \left(cos(\gamma) - \frac{1}{2\,cos(\gamma)}\right) F_{p,s} - \frac{F_{p,b}}{2} tan(\gamma) \\ F_{A,z} &= F_{p,s}\, sin(\gamma) + \frac{F_{p,b}}{2} \\ F_{B,y} &= \left(\frac{1}{2\,cos(\gamma)} - cos(\gamma)\right) F_{p,s} - \frac{F_{p,b}}{2} tan(\gamma) \\ F_{B,z} &= -F_{p,s}\, sin(\gamma) + \frac{F_{p,b}}{2} \end{aligned} \qquad \textbf{8.50}$$

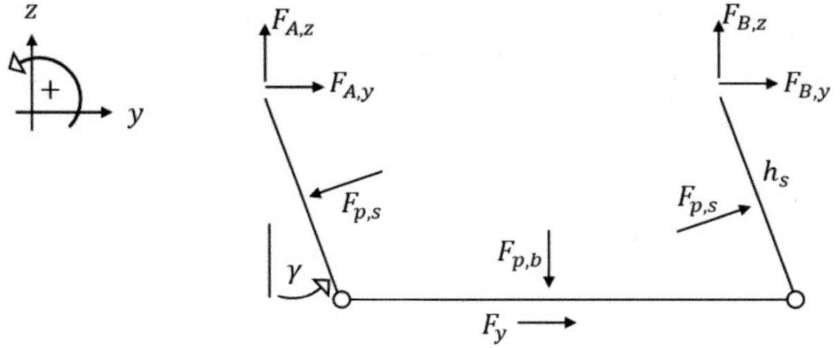

Figure 8.6: Shear modulus due to inflation pressure

From the lateral force equilibrium the following equation can be derived:

$$\sum F_y = F_y - F_{p,b}\,tan(\gamma) = 0 \qquad \textbf{8.51}$$

and, subsequently:

$$p = \frac{F_y / A_{p,xy}}{tan(\gamma)} = G_{yp} \qquad \textbf{8.52}$$

In order to determine the shear modulus G_{ys} of the sidewalls, the sidewall is modelled as a bar (see Figure 8.7). The sidewall is fixed at the top with a clamp (Y_B, Z_B, M_B, all displacements are blocked) and at the bottom with a translating joint (M_A is blocked, the only degree of freedom is in lateral direction y) and the vertical force is given. The sidewall is in static equilibrium.

The system has more degrees of boundaries (DoB) than degrees of freedom (DoF): the unknown torque X acting at the boundaries can be calculated according to the principle of virtual work. To this purpose, it is necessary to distinguish between real system, which includes all forces as shown in Figure 8.7, and virtual system, which includes only the unknown torque. In this case, torque M_A is chosen as unknown quantity. Axial forces N, shear forces T, as well as torques M of real and virtual system are enlisted in Table 8.8.

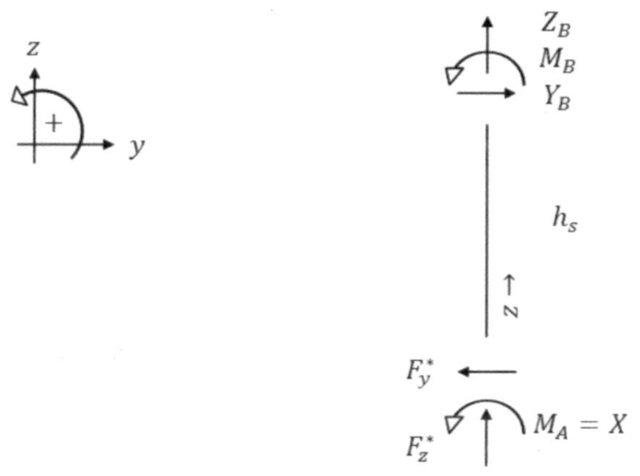

Figure 8.7: Modelling of the sidewall

It should be noticed, that the value of force F_y^* can be set arbitrarily, as it does not influence the solution (see equation 8.58). The principle of virtual work is:

$$\int \frac{N_{real} N_{virt}}{E_s A_{s,xy}} dz + \int \frac{T_{real} T_{virt}}{G_s A_{s,xy}} dz + \int \frac{M_{real} M_{virt}}{E_s I_{s,yx}} dz = 0 \quad \textbf{8.53}$$

In our case the contribution of the axial work is zero and the contribution of the shear force is negligible compared to the contribution of the work done by the torque:

$$\int \frac{M_{real} M_{virt}}{E_s I_{s,yx}} dz = 0 \quad \textbf{8.54}$$

The unknown torque X is then:

$$X = \frac{1}{2} \cdot F_y^* \cdot z \quad \textbf{8.55}$$

The lateral deformation δ_s associated to the sidewall is:

$$E_s I_{s,yx} \frac{\partial^2 \delta_s(z)}{\partial z^2} = M_{real} \quad \textbf{8.56}$$

Table 8.8: Forces and torques of real and virtual system

Real System	Virtual System
$X = x$	$X = 1\ Nm$
$F_z^* = F_z/2$	$F_z^* = 0$
$F_y^* = 4000\ N$	$F_y^* = 0$
$N_{real} = -F_z^*$	$N_{virt} = 0$
$T_{real} = F_y^*$	$T_{virt} = 0$
$M_{real} = -X + F_y^* z$	$M_{virt} = -1\ Nm$

Imposing the two boundary conditions:

$$\delta_s(h_s) = 0 \qquad \delta_s'(0) = 0 \tag{8.57}$$

the expression of the lateral deflection of the sidewall is (positive if directed as F_y^*):

$$\delta_s(z) = -\frac{F_y^*}{12E_sI_{s,yx}}z^3 + \frac{F_y^*}{12E_sI_{s,yx}}h_s^3 \tag{8.58}$$

Subsequently, the shear modulus of the sidewall is:

$$G_{ys} = f_{G_{ys}} \cdot \left.\frac{F_y^*/A_{s,xy}}{\delta_s/h_s}\right|_{z=0} = f_{G_{ys}} \cdot \frac{12E_sI_{s,yx}}{A_{s,xy}h_s^2} \tag{8.59}$$

8.1.3.8 Shear Modulus of Belt and Tread

The shear modulus G_{y2} of the belt and the tread is the sum of two shear moduli "in series":

$$G_{y2} = \left(\left(G_{yb}\frac{A_{c,xy}}{h_b}\right)^{-1} + \left(G_{yt}\frac{A_{c,xy}}{h_t}\right)^{-1}\right)^{-1} \frac{h_{Gy2}^*}{A_{Gy2}^*} \tag{8.60}$$

where G_{yb} is the shear modulus of the belt and G_{yt} the shear modulus of the tread. Height h_{Gy2}^* and area A_{Gy2}^* are assumed to be (see section 8.1.3.5):

$$h_{Gy2}^{*} = 2 \cdot \left(\left(\frac{G_{yb}}{h_b} \right)^{-1} + \left(\frac{G_{yt}}{h_t} \right)^{-1} \right) \cdot \left(G_{yb}^{-1} + G_{yt}^{-1} \right)^{-1} \quad \mathbf{8.61}$$

$$A_{Gy2}^{*} = \left(\left(G_{yb} A_{b,xy} \right)^{-1} + \left(G_{yt} A_{t,xy} \right)^{-1} \right)^{-1} \cdot \left(G_{yb}^{-1} + G_{yt}^{-1} \right) \quad \mathbf{8.62}$$

and shear modulus of belt and tread are estimated as:

$$G_{yb} = \frac{1}{3} E_b \quad \mathbf{8.63}$$

$$G_{yt} = \frac{1}{3} E_t \quad \mathbf{8.64}$$

considering that Poisson's ratio ν is non-negative and generally not bigger than 0.5 (see equation 8.34).

8.1.3.9 Lateral Stiffness of Sidewalls and Inflation Pressure

Lateral stiffness k_{y1} (see Figure 8.11) of sidewalls and due to inflation pressure is assumed to be constant with frequency and amplitude (small displacements):

$$k_{y1} = G_{y1} \frac{A_{Gy1}^{*}}{h_{Gy1}^{*}} \quad \mathbf{8.65}$$

where G_{y1}, A_{Gy1}^{*} and h_{Gy1}^{*} are defined in section 8.1.3.7.

8.1.3.10 Lateral Stiffness of Belt and Tread

Lateral stiffness k_{y2} (see Figure 8.11) of belt and tread is assumed to be constant with frequency and amplitude (small displacements):

$$k_{y2} = G_{y2} \frac{A_{Gy2}^{*}}{h_{Gy2}^{*}} \quad \mathbf{8.66}$$

where G_{y2}, A_{Gy2}^{*} and h_{Gy2}^{*} are defined in section 8.1.3.8.

8.1.3.11 Damping

All damping coefficient are assumed to be constant with frequency and amplitude (small displacements). Their values are adjusted to measurements of vertical force spectra on a tyre test bench.

Concerning vertical damping of the tyre (see Figure 8.9), the contribution of the inflation pressure is assumed to be negligible. Then, the damping between belt and rim is assumed to be the sum in parallel of the damping of the two sidewalls:

$$d_{z1} = f_{d_z} \cdot 2d_s \qquad \textbf{8.67}$$

where f_{d_z} is a form factor taking into account the tyre structure. Concerning the damping between contact area and belt, the damping is assumed to be the sum in series of the damping of belt and tread:

$$d_{z2} = f_{d_z} \cdot (d_b^{-1} + d_t^{-1})^{-1} \qquad \textbf{8.68}$$

Concerning lateral damping of the tyre (see Figure 8.11), damping d_{y1} of sidewall and due to inflation pressure and damping d_{y2} of belt and tread are estimated by the form factor f_{d_y}:

$$d_{y1} = f_{d_y} d_{z1} \qquad \textbf{8.69}$$

$$d_{y2} = f_{d_y} d_{z2} \qquad \textbf{8.70}$$

8.1.3.12 Mass

The tyre model includes three masses (see Figure 8.9). The mass of the air in the cavity is:

$$m_a = \rho_a \cdot 2\pi \left(R - \frac{h_s}{2}\right) \cdot w_n h_s \qquad \textbf{8.71}$$

where ρ_a is the density of the air ($1.2\ kg/m^3$).

8.1.4 Static Equilibrium

Given a vertical force F_z, static equilibrium is reached if the following conditions are satisfied:

$$F_z = \int_{\tilde{z}_r}^{\tilde{z}_b} k_{z1}\, dz_b = \int_{\tilde{z}_r}^{\tilde{z}_b} 2 f_{k_s} \frac{E_s A_{s,xy}}{h_s} dz_b + f_{k_s} p A_{p,xy} \qquad \textbf{8.72}$$

$$F_z = \int_{\tilde{z}_b}^{\tilde{z}} k_{z2}\, dz = \int_{\tilde{z}_b}^{\tilde{z}} \left(\left(\frac{E_b A_{c,xy}}{h_b} \right)^{-1} + \left(\frac{E_t A_{c,xy}}{h_t} \right)^{-1} \right)^{-1} dz \qquad \textbf{8.73}$$

where $\tilde{z}_r$, $\tilde{z}_b$ and $\tilde{z}$ are the vertical displacements of rim, belt and road surface at a static tyre load. For static equilibrium conditions, it is possible to linearize all geometrical and material properties.

8.2 Ride Comfort and Interior Noise

In this section, first, the equations of motion for vertical and lateral dynamics are presented. Then, the frequency response functions (FRFs) used to characterize ride comfort and interior noise of the tyre are derived.

8.2.1 Three Mass Model for Vertical Direction

The model for vertical direction has three degrees of freedom given by the vertical displacements of the masses of rim (which includes the mass of the bead), air in the cavity and tyre. The masses are linked by Kelvin models (a spring and a damper connected in parallel). Often, rubber is modelled using a Gehmann model, which has an additional spring in series to the damping element (Harris & Crede, 1976).

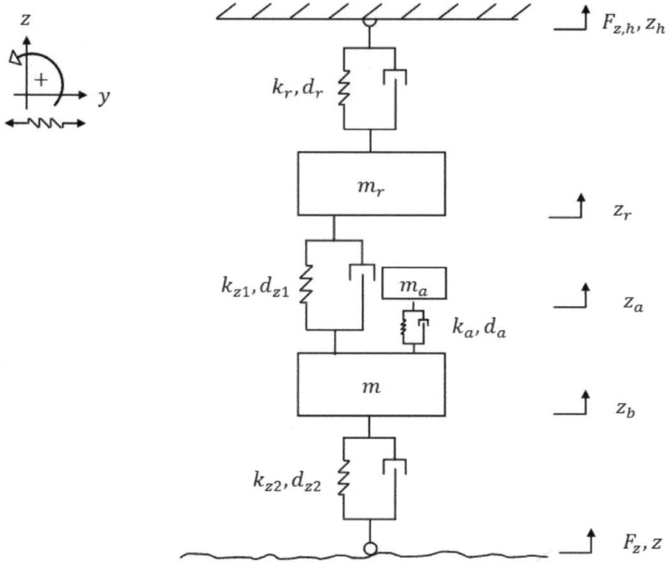

Figure 8.9: Three mass model (vertical)

In contrast to the Kelvin model, the stiffness increases with frequency and the damping decreases with frequency. As the increment of stiffness is very small, the main difference between the two models is given only by the damping characteristics (Mitschke & Wallentowitz, Dynamik der Kraftfahrzeuge, 2003). For the modelling the Kelvin model is preferred, as it is the best compromise between simplicity and precision. The model is represented in Figure 8.9 where m represents the mass of the tyre, m_a of the air in the cavity and m_r of the rim (all other TDPs are enlisted in Table 8.1). Positive are counter-clockwise rotations, upward displacements and stretched springs and dampers. The vector $\bar{z}$ containing the independent variables is:

$$\bar{z} = \begin{Bmatrix} z_r \\ z_a \\ z_b \\ \hdashline z_h \\ z \end{Bmatrix} = \begin{Bmatrix} \bar{z}_f \\ \bar{z}_b \end{Bmatrix} \qquad \textbf{8.74}$$

where z_r, z_a, and z_b are independent variables describing the deflection of respectively rim, air and belt (the three DoF $\bar{z}_f$, where the subscripted character f refers to "free"); z_h and z are independent variables describing the displacements of hub and road (the two DoB $\bar{z}_b$, where the subscripted character b refers to "bounded"). All independent variables describe the position of a mass or a bond relative to the configuration of non-loaded tyre.

8.2.1.1 Equations of Motion

The equations of motion for vertical direction are derived using the Lagrangian mechanics:

$$\frac{d}{dt}\frac{\partial E_c}{\partial \dot{z}_i} - \frac{\partial E_c}{\partial z_i} + \frac{\partial D}{\partial \dot{z}_i} + \frac{\partial V}{\partial z_i} = \frac{\delta W_i}{\delta z_i} \qquad \mathbf{8.75}$$

The kinetic energy of the system is:

$$E_c = \frac{1}{2}m_r\dot{z}_r^2 + \frac{1}{2}m_a\dot{z}_a^2 + \frac{1}{2}m\dot{z}_b^2 \qquad \mathbf{8.76}$$

The damping dissipation of the system is:

$$D = \frac{1}{2}d_r(\dot{z}_h - \dot{z}_r)^2 + \frac{1}{2}d_{z1}(\dot{z}_r - \dot{z}_b)^2 + \frac{1}{2}d_a(\dot{z}_a - \dot{z}_b)^2 + \frac{1}{2}d_{z2}(\dot{z}_b - \dot{z})^2 \qquad \mathbf{8.77}$$

The potential energy of the system is:

$$V = \frac{1}{2}k_r(z_h - z_r)^2 + \frac{1}{2}\tilde{k}_{z1}(z_r - z_b)^2 + \frac{1}{2}k_a(z_a - z_b)^2 + \frac{1}{2}\tilde{k}_{z2}(z_b - z)^2 \qquad \mathbf{8.78}$$

where $\tilde{k}_i$ are the stiffnesses linearized around static load (otherwise a function of deflection). The work done by forces acting on the system is:

$$W = F_{z,h}z_h + F_z z \qquad \mathbf{8.79}$$

The equations of motion of the linearized systems are:

$$[M]\ddot{\bar{z}} + [D]\dot{\bar{z}} + [K]\bar{z} = \bar{F} \tag{8.80}$$

where the vector $\bar{z}$ contains the independent variables.
The mass matrix (symmetric and defined positive) is:

$$[M] = \left[\begin{array}{ccc|cc} m_r & & & & \\ & m_a & & & \\ & & m & & \\ \hline & & & 0 & \\ & & & & 0 \end{array}\right] = \begin{bmatrix} M_{ff} & M_{fb} \\ M_{bf} & M_{bb} \end{bmatrix} \tag{8.81}$$

The damping matrix is (symmetric):

$$[D] = \left[\begin{array}{ccc|cc} d_r + d_{z1} & 0 & -d_{z1} & -d_r & 0 \\ 0 & d_a & -d_a & 0 & 0 \\ -d_{z1} & -d_a & d_{z1} + d_a + d_{z2} & 0 & -d_{z2} \\ \hline -d_r & 0 & 0 & d_r & 0 \\ 0 & 0 & -d_{z2} & 0 & d_{z2} \end{array}\right] = \begin{bmatrix} D_{ff} & D_{fb} \\ D_{bf} & D_{bb} \end{bmatrix} \tag{8.82}$$

The stiffness matrix is (symmetric):

$$[K] = \left[\begin{array}{ccc|cc} k_r + \tilde{k}_{z1} & 0 & -\tilde{k}_{z1} & -k_r & 0 \\ 0 & k_a & -k_a & 0 & 0 \\ -\tilde{k}_{z1} & -k_a & \tilde{k}_{z1} + k_a + \tilde{k}_{z2} & 0 & -\tilde{k}_{z2} \\ \hline -k_r & 0 & 0 & k_r & 0 \\ 0 & 0 & -\tilde{k}_{z2} & 0 & \tilde{k}_{z2} \end{array}\right] = \begin{bmatrix} K_{ff} & K_{fb} \\ K_{bf} & K_{bb} \end{bmatrix} \tag{8.83}$$

The force vector is:

$$\bar{F} = \left\{\begin{array}{c} 0 \\ 0 \\ 0 \\ \hline F_{z,h} \\ F_z \end{array}\right\} = \begin{Bmatrix} \bar{F}_f \\ \bar{F}_b \end{Bmatrix} \tag{8.84}$$

8.2.1.2 Natural Frequencies of the System

The natural frequencies ω_r, ω_a and ω_b are determined assuming:

$$[M_{ff}]\ddot{\bar{z}}_f + [K_{ff}]\bar{z}_f = \bar{0} \tag{8.85}$$

$$(-[\omega]^2[M_{ff}] + [K_{ff}])\bar{z}_{f,0}e^{i\omega t} = \bar{0} \tag{8.86}$$

$$[\omega] = \begin{bmatrix} \omega_r & & \\ & \omega_a & \\ & & \omega_b \end{bmatrix} = \sqrt{[M_{ff}]^{-1}[K_{ff}]} \tag{8.87}$$

The related eigenmodes $z_{f,0,r}$, $z_{f,0,a}$ and $z_{f,0,b}$ of the system are determined by:

$$(-\omega_i{}^2[M_{ff}] + [K_{ff}])z_{f,0,i} = \bar{0} \qquad i \in [r, a, b] \tag{8.88}$$

8.2.1.3 Frequency Response Function

The frequency response function (FRF) of the vertical dynamics is derived under the assumption of fixed wheel hub ($z_h = 0$). Subsequently, $\bar{z}_b$ is reduced to the second row (2) of its first column (1):

$$\tilde{\bar{z}}_b = \bar{z}_b(2,1) = \{z\} \tag{8.89}$$

and also the matrices $[M_{fb}]$, $[D_{fb}]$, $[K_{fb}]$ are reduced to the first three rows (1: 3) of their second column (2), e.g.:

$$[\widetilde{M}_{fb}] = [M_{fb}(1{:}3, 2)] \tag{8.90}$$

Taking the first three equations of the linearized system (see equation 8.80), the FRFs of rim, air and tyre to an excitation of the ground are:

$$\begin{aligned}[M_{ff}]\ddot{\bar{z}}_f + [D_{ff}]\dot{\bar{z}}_f + [K_{ff}]\bar{z}_f \\ = \bar{F}_f - \left([\widetilde{M}_{fb}]\ddot{\tilde{\bar{z}}}_b + [\widetilde{D}_{fb}]\dot{\tilde{\bar{z}}}_b + [\widetilde{K}_{fb}]\tilde{\bar{z}}_b\right)\end{aligned} \tag{8.91}$$

$$\left(-\omega^2[M_{ff}] + i\omega[D_{ff}] + [K_{ff}]\right)\bar{z}_{f,0}e^{i\omega t} \\ = -\left(-\omega^2[\widetilde{M}_{fb}] + i\omega[\widetilde{D}_{fb}] + [\widetilde{K}_{fb}]\right)\bar{\bar{z}}_{b,0}e^{i\omega t} \quad 8.92$$

$$\bar{H} = \frac{\bar{z}_{f,0}}{\bar{\bar{z}}_{b,0}} = -\left(-\omega^2[M_{ff}] + i\omega[D_{ff}] + [K_{ff}]\right)^{-1}\left(-\omega^2[\widetilde{M}_{fb}] \\ + i\omega[\widetilde{D}_{fb}] + [\widetilde{K}_{fb}]\right) \quad 8.93$$

The vector $\bar{H}$ contains the three FRFs:

$$\bar{H} = \begin{Bmatrix} H_{z,r}(\omega) \\ H_{z,a}(\omega) \\ H_{z,b}(\omega) \end{Bmatrix} = \begin{Bmatrix} \dfrac{z_{r,0}(\omega)}{z_0(\omega)} \\ \dfrac{z_{a,0}(\omega)}{z_0(\omega)} \\ \dfrac{z_{b,0}(\omega)}{z_0(\omega)} \end{Bmatrix} \quad 8.94$$

Taking the fourth equation of the linearized system (see equation 8.80), the force spectrum at the wheel hub is calculated:

$$-d_r\dot{z}_r - k_r z_r = F_{z,h} \quad 8.95$$

$$-(i\omega d_r + k_r)z_{r,0}e^{i\omega t} = F_{z,h,0}e^{i\omega t} \quad 8.96$$

$$\frac{F_{z,h,0}(\omega)}{z_{r,0}(\omega)} = -(i\omega d_r + k_r) \quad 8.97$$

Using the fifth equation, the force spectrum in the contact path is determined:

$$d_{z2}(\dot{z} - \dot{z}_b) + k_{z2}(z - z_b) = F_z \quad 8.98$$

$$(i\omega d_{z2} + k_{z2})(z_0 - z_{b,0})e^{i\omega t} = F_{z,0}e^{i\omega t} \quad 8.99$$

$$\frac{F_{z,0}(\omega)}{z_0(\omega)} = (i\omega d_{z2} + k_{z2})(1 - H_{z,b}) \quad 8.100$$

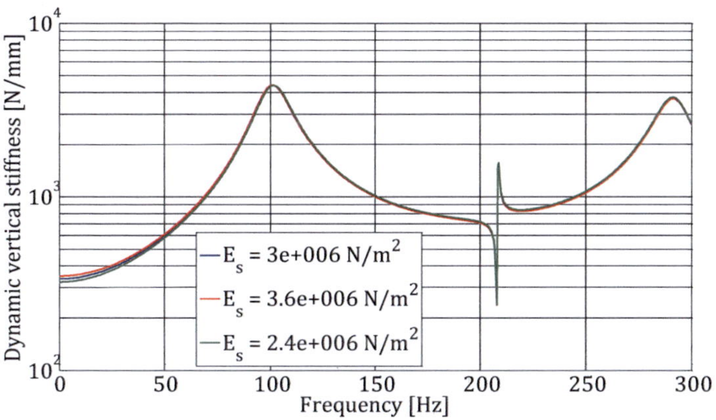

Figure 8.10: Variation of the dynamic vertical stiffness (physical tyre model)

If the system is linear, the effects can be superposed. Subsequently, the FRF between road excitation vertical displacement and force at the wheel hub is:

$$\begin{aligned} H_{z,1}(\omega) &= \frac{F_{z,h,0}(\omega)}{z_0(\omega)} = \frac{F_{z,h,0}(\omega)}{z_{r,0}(\omega)} \cdot \frac{z_{r,0}(\omega)}{z_0(\omega)} \\ &= -(i\omega d_r + k_r) \cdot H_{z,r} \end{aligned} \qquad \textbf{8.101}$$

Similarly, the FRF between the road excitation induced vertical force and force at the wheel hub is:

$$\begin{aligned} H_{z,2}(\omega) &= \frac{F_{z,h,0}(\omega)}{F_{z,0}(\omega)} = \frac{F_{z,h,0}(\omega)}{z_0(\omega)} \cdot \frac{z_0(\omega)}{F_{z,0}(\omega)} \\ &= H_{z,1} \cdot \left((i\omega d_{z2} + k_{z2})(1 - H_{z,b}) \right)^{-1} \end{aligned} \qquad \textbf{8.102}$$

In Figure 8.10 the variation of the dynamic vertical stiffness (see equation 8.101) due to the elastic modulus of the sidewall is exemplarily shown for a 205/55 R16 run-flat tyre (see Table 8.1).

8.2.2 Two Mass Model for Lateral Direction

The model for lateral direction is very similar to the model in vertical direction, but is has only two degrees of freedom, as the air cavity does not influence lateral dynamics.

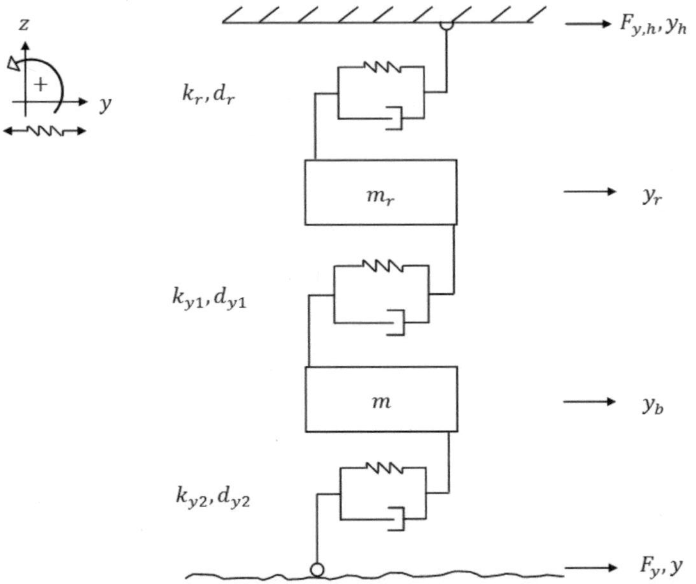

Figure 8.11: Two mass model (lateral)

The model is represented in Figure 8.11. Positive are counter-clockwise rotations, displacements to the right and stretched springs and dampers.

The vector $\bar{y}$ containing the independent variables is:

$$\bar{y} = \begin{Bmatrix} y_r \\ y_b \\ \hline y_h \\ y \end{Bmatrix} = \begin{Bmatrix} \bar{y}_f \\ \bar{y}_b \end{Bmatrix} \qquad \textbf{8.103}$$

where y_r and y_b are independent variables describing the deflection of respectively rim and belt (the two DoF $\bar{y}_f$); y_h and y are independent variables describing the displacements of hub and road (the two DoB $\bar{y}_b$). All independent variables describe the position of a mass or a bond relative to the configuration of non-loaded tyre.

8.2.2.1 Equations of Motion

The equations of motion for lateral direction are derived using the Lagrangian mechanics:

$$\frac{d}{dt}\frac{\partial E_c}{\partial \dot{y}_i} - \frac{\partial E_c}{\partial y_i} + \frac{\partial D}{\partial \dot{y}_i} + \frac{\partial V}{\partial y_i} = \frac{\delta W_i}{\delta y_i} \qquad \textbf{8.104}$$

The equations of motion of the linearized systems are:

$$[M]\ddot{\bar{y}} + [D]\dot{\bar{y}} + [K]\bar{y} = \bar{F} \qquad \textbf{8.105}$$

The mass matrix (symmetric and defined positive) is:

$$[M] = \left[\begin{array}{cc|cc} m_r & & & \\ & m & & \\ \hline & & 0 & \\ & & & 0 \end{array}\right] = \begin{bmatrix} M_{ff} & M_{fb} \\ M_{bf} & M_{bb} \end{bmatrix} \qquad \textbf{8.106}$$

The damping matrix is (symmetric):

$$[D] = \left[\begin{array}{cc|cc} d_r + \tilde{d}_{y1} & -\tilde{d}_{y1} & -d_r & 0 \\ -\tilde{d}_{y1} & \tilde{d}_{y1} + \tilde{d}_{y2} & 0 & -\tilde{d}_{y2} \\ \hline -d_r & 0 & d_r & 0 \\ 0 & -\tilde{d}_{y2} & 0 & \tilde{d}_{y2} \end{array}\right] = \begin{bmatrix} D_{ff} & D_{fb} \\ D_{bf} & D_{bb} \end{bmatrix} \qquad \textbf{8.107}$$

The stiffness matrix (symmetric) is:

$$[K] = \left[\begin{array}{cc|cc} k_r + \tilde{k}_{y1} & -\tilde{k}_{y1} & -k_r & 0 \\ -\tilde{k}_{y1} & \tilde{k}_{y1} + \tilde{k}_{y2} & 0 & -\tilde{k}_{y2} \\ \hline -k_r & 0 & k_r & 0 \\ 0 & -\tilde{k}_{y2} & 0 & \tilde{k}_{y2} \end{array}\right] = \begin{bmatrix} K_{ff} & K_{fb} \\ K_{bf} & K_{bb} \end{bmatrix} \qquad \textbf{8.108}$$

The force vector is:

$$\bar{F} = \begin{Bmatrix} 0 \\ 0 \\ F_{y,h} \\ F_y \end{Bmatrix} = \begin{Bmatrix} \bar{F}_f \\ \bar{F}_b \end{Bmatrix} \qquad \textbf{8.109}$$

8.2.2.2 Natural Frequencies of the System

As described in section 8.2.1.2 the natural frequencies ω_r and ω_m and their related eigenmodes $y_{f,0,r}$ and $y_{f,0,b}$ are:

$$[\omega] = \begin{bmatrix} \omega_r & \\ & \omega_b \end{bmatrix} = \sqrt{\left[M_{ff}\right]^{-1}\left[K_{ff}\right]} \qquad \textbf{8.110}$$

$$\left(-\omega_i{}^2\left[M_{ff}\right] + \left[K_{ff}\right]\right)y_{f,0,i} = \bar{0} \qquad i \in [r, a, b] \qquad \textbf{8.111}$$

8.2.2.3 Frequency Response Function

As described in section 8.2.1.3 the frequency response function (FRF) between road excitation and force at the wheel hub is:

$$\begin{aligned} H_{y,1}(\omega) &= \frac{F_{y,h,0}(\omega)}{y_0(\omega)} = \frac{F_{y,h,0}(\omega)}{y_{r,0}(\omega)} \cdot \frac{y_{r,0}(\omega)}{y_0(\omega)} \\ &= -(i\omega d_r + k_r) \cdot H_{y,r} \end{aligned} \qquad \textbf{8.112}$$

Similarly, the FRF between road excitation induced lateral force and force at the wheel hub is:

$$\begin{aligned} H_{y,2}(\omega) &= \frac{F_{y,h,0}(\omega)}{F_{y,0}(\omega)} = \frac{F_{y,h,0}(\omega)}{y_0(\omega)} \cdot \frac{y_0(\omega)}{F_{y,0}(\omega)} \\ &= H_{h,1} \cdot \left(\left(i\omega d_{y2} + k_{y2}\right)\left(1 - H_{y,b}\right)\right)^{-1} \end{aligned} \qquad \textbf{8.113}$$

In Figure 8.12 the variation of the dynamic lateral stiffness (see equation 8.112) due to the elastic modulus of the sidewall is exemplarily shown for a 205/55 R16 run-flat tyre (see Table 8.1).

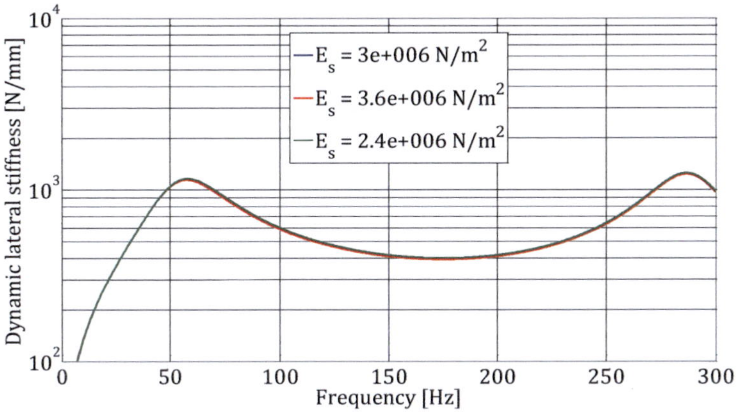

Figure 8.12: Variation of the dynamic lateral stiffness (physical tyre model)

8.3 Lateral Dynamics

In this section, the brush model, used to describe the quasi-static lateral force, is presented.

8.3.1 Brush Model

In Figure 8.13 a classical representation of the brush model is given. The kinematic shear strain of the tread bars is represented by γ_k.

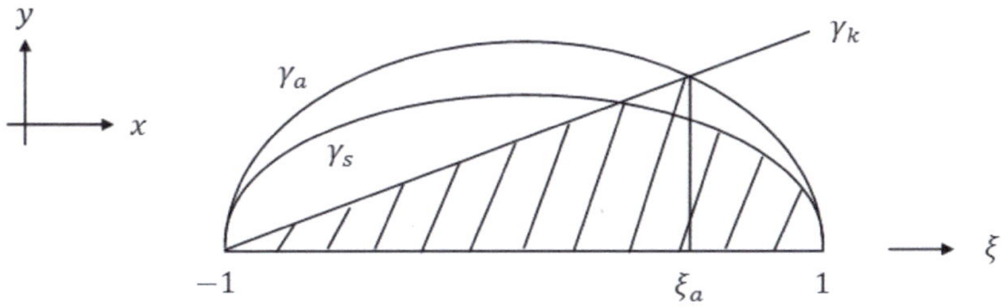

Figure 8.13: Brush model for lateral dynamics

At the entrance of the contact area ($\xi = -1$) the kinematic shear strain is zero, as the tread bars are non-deformed; then, it increases linearly. From the tread entrance to $\xi = \xi_a$ the tread bars are subjected to adhesion conditions; from $\xi = \xi_a$ to the exit of the contact area, instead, the tread bars are subjected to slip conditions.

Assuming that the distributions of contact pressure and friction coefficients are constant over tyre width, lateral force F_y is:

$$\begin{aligned} F_y(\xi_a) &= \frac{1}{2}\int_{-1}^{1} \tau(\xi,\xi_a)\cdot A^*_{Gy}\, d\xi \\ &= \frac{1}{2}\int_{-1}^{1} G_y\cdot\gamma(\xi,\xi_a)\cdot A^*_{Gy}\, d\xi \\ &\cong G_y\cdot\bar{\gamma}\cdot A^*_{Gy} \end{aligned} \qquad \mathbf{8.114}$$

where G_y is the total shear modulus, A^*_{Gy} the area on which the shear modulus works, $\xi \in [-1,1]$ the normalised coordinate describing the longitudinal position along the contact path and $\xi_a \in [-1,1]$ the critical normalised coordinate. The related lateral slip angle is:

$$\alpha(\xi_a) = \tan^{-1}\left(\frac{h^*_{Gy}}{l_c}\frac{\gamma_a(\xi_a)}{(\xi_a+1)/2}\right) \qquad \mathbf{8.115}$$

where h^*_{Gy} is the tyre height on which the shear modulus works, l_c the length of the contact path and γ_a the shear strain. Shear strain $\bar{\gamma}$ is the composed of the integrals of kinematic shear strain γ_k and of sliding shear strain γ_s:

$$\bar{\gamma} = \frac{1}{2}\int_{-1}^{\xi_a} \gamma_k(\xi,\xi_a)\, d\xi + \frac{1}{2}\int_{\xi_a}^{1} \gamma_s(\xi)\, d\xi \qquad \mathbf{8.116}$$

Where kinematic shear strain γ_k is a function of shear strain γ_a for adhesion conditions:

$$\gamma_k(\xi,\xi_a) = \frac{\gamma_a(\xi_a)}{(\xi_a+1)/2}\frac{\xi+1}{2} \qquad \textbf{8.117}$$

Shear strain γ_a for adhesion conditions is:

$$\gamma_a(\xi) = \mu_a \frac{p_c}{G_y} \qquad \textbf{8.118}$$

where μ_a is the adhesion coefficient distribution and p_c the longitudinal pressure distribution in the contact area. Similarly, shear strain γ_s for sliding conditions is:

$$\gamma_s(\xi) = \mu_s \frac{p_c}{G_y} \qquad \textbf{8.119}$$

where μ_s is the sliding coefficient distribution. The ground pressure distribution p_c can be modelled empirically:

$$p_c(\xi) = \frac{F_z}{A_{c,xy}} \cdot \frac{q_1+1}{q_1}(1-|\xi|^{q_1}) \qquad \textbf{8.120}$$

where F_z is the tyre load, $A_{c,xy}$ the contact area and q_1 an exponent for the pressure distribution with values between 3 and 8 (Ammon, Gnadler, Mäckle, & Unrau, 2004).

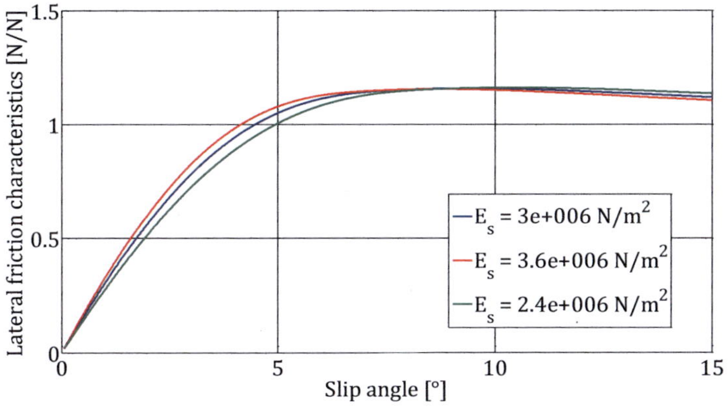

Figure 8.14: Variation of the lateral friction characteristics (physical tyre model)

The adhesion coefficient distribution can be modelled empirically:

$$\mu_a(\xi) = \mu_{a,ref} \left(\frac{p_c}{p_{ref}} \right)^{-q_2} \qquad \textbf{8.121}$$

where $\mu_{a,ref}$ is the reference adhesion coefficient, p_{ref} is the reference ground pressure set equal to $2.5 \cdot 10^5$ (Ammon, Gnadler, Mäckle, & Unrau, 2004), q_2 an exponent for the friction coefficient distribution with value 0.3 (Ammon, Gnadler, Mäckle, & Unrau, 2004). Similarly, the distribution of the sliding coefficient is:

$$\mu_s(\xi) = \mu_{s,ref} \left(\frac{p_c}{p_{ref}} \right)^{-q_2} \qquad \textbf{8.122}$$

where $\mu_{s,ref}$ is the reference sliding coefficient linked to the reference adhesion coefficient $\mu_{a,ref}$ trough the factor q_3:

$$\mu_{s,ref} = q_3 \cdot \mu_{a,ref} \qquad \textbf{8.123}$$

In Figure 8.14 the variation of the lateral friction characteristics (see equation 8.114) due to the elastic modulus of the sidewall is exemplarily shown for a 205/55 R16 run-flat tyre (see Table 8.1).

8.4 Power Loss

In this section, the hysteretic behaviour of a visco-elastic material, used to describe the power loss of a rolling tyre, is presented.

8.4.1 Hysteresis Model

There are several formulations for the power loss of a tyre, amongst others the product of rolling resistance force $F_{r,0}$ and trajectory velocity v_x:

$$P_{loss} = F_{r,0} \cdot v_x \qquad \textbf{8.124}$$

where rolling resistance force $F_{r,0}$ can be related to the energy loss per cycle E_{loss}:

$$F_{r,0} = \frac{1}{2\pi R} E_{loss} \qquad \textbf{8.125}$$

The energy loss per cycle E_{loss} of a tyre is supposed to be the sum of energy loss $E_{loss,V}$ caused by hysteresis under a sinusoidal vertical displacement of the tyre and energy loss $E_{loss,B}$ caused by hysteresis due to the variation of curvature of sidewall, belt and tread when entering the contact area:

$$E_{loss} = E_{loss,V} + E_{loss,B} \qquad \textbf{8.126}$$

Imposing a sinusoidal strain on a visco-elastic material will result in a sinusoidal stress, which generally is not in phase with the imposed deformation. The material will then exhibit a hysteretic behaviour as shown in Figure 8.15 (Clark, 1981). Stress σ and strain ε over time t and angular frequency ω are defined as:

$$\begin{aligned}\sigma &= \sigma_0 \sin(\omega t + \delta) \\ &= \sigma_0 \cos(\delta) \sin(\omega t) + \sigma_0 \sin(\delta) \cos(\omega t) \\ &= E' \varepsilon_0 \sin(\omega t) + E'' \varepsilon_0 \cos(\omega t)\end{aligned} \qquad \textbf{8.127}$$

$$\varepsilon = \varepsilon_0 \sin(\omega t) \qquad \textbf{8.128}$$

where E' is the storage modulus, E'' the loss modulus, σ_0 and ε_0 the stress and strain amplitude.
The loss tangent related to the phase angle δ is defined as:

$$\tan(\delta) = \frac{E''}{E'} \qquad \textbf{8.129}$$

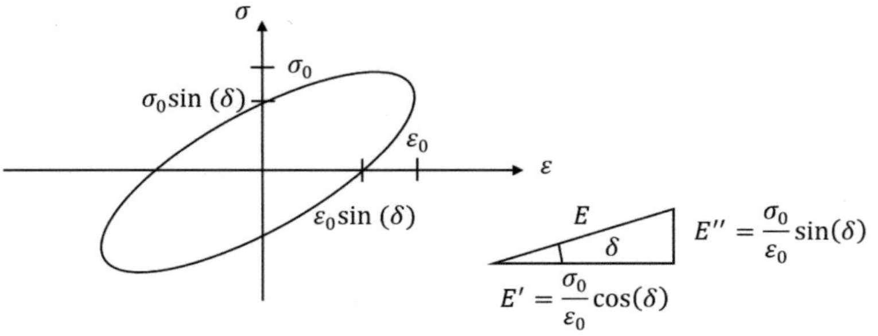

Figure 8.15: Representation of the hysteretic behaviour

The energy loss per volume and cycle caused by hysteresis under a sinusoidal vertical displacement of the tyre is (Clark, 1981):

$$\frac{dE_{loss,V_i}}{dV_i} = \int_0^{2\pi/\omega} \sigma d\varepsilon = \int_0^{2\pi/\omega} \sigma \frac{d\varepsilon}{dt} dt = \pi E'' \varepsilon_0^2 \qquad \textbf{8.130}$$

Subsequently, the energy loss of the tyre due to hysteresis is:

$$\begin{aligned} E_{loss,V} &= \sum_i \left[\int_{V_i} \pi \sigma_{i,0}\, \varepsilon_{i,0}\, sin(\delta_i)\, dV_i \right] \\ &\cong \sum_i \left[\pi \sigma_{i,0}\, \varepsilon_{i,0}\, sin(\delta_i)\, V_i \right] \end{aligned} \qquad i \in [s, b, t] \qquad \textbf{8.131}$$

where V the volume of tyre tread, belt and sidewalls. The stress amplitude of sidewall, belt and tread are respectively approximated by:

$$\sigma_{s,0} = \frac{(F_z - pA_{p,xy})/2}{A_{s,xy}} \qquad \textbf{8.132}$$

$$\sigma_{b,0} = \frac{F_z}{A_{c,xy}} \qquad \textbf{8.133}$$

$$\sigma_{t,0} = \frac{F_z}{A_{c,xy}} \qquad \textbf{8.134}$$

The strain amplitude of sidewall, belt and tread are respectively approximated by:

$$\varepsilon_{s,0} = \frac{|z_r - z_b|}{h_s} \qquad \textbf{8.135}$$

$$\varepsilon_{b,0} = \frac{|z_b - z|}{h_b} \qquad \textbf{8.136}$$

$$\varepsilon_{t,0} = \frac{|z_b - z|}{h_t} \qquad \textbf{8.137}$$

The loss tangent can be related to stiffness and damping characteristics of a model used to represent the dynamic behaviour of a rubber element modelled through a Kelvin model (Mitschke & Wallentowitz, 2003). Subsequently, the phase angles between stress and strain of sidewall, belt and tread are:

$$\delta_i = tan^{-1}\left(\frac{d_i}{k_i}\omega\right) \qquad i \in [s, b, t] \qquad \textbf{8.138}$$

where d_i are the damping, k_i the stiffnesses and ω the angular frequency. The angular frequency is related to the trajectory velocity:

$$\omega = \frac{v_x}{R} \qquad \textbf{8.139}$$

The volume of tyre tread, belt and sidewalls is approximated by:

$$V_i = u_i \cdot w_i h_\mathrm{i} \qquad i \in [s, b, t] \qquad \textbf{8.140}$$

where the average circumferences u_i of tyre tread, belt and sidewall are evaluated as follows:

$$u_s = 2\pi\left(R - \left(h_t + h_b + \frac{h_s}{2}\right)\right) \qquad \textbf{8.141}$$

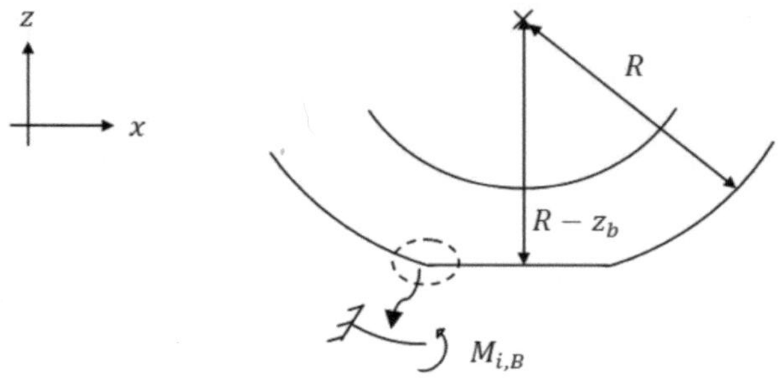

Figure 8.16: Modelling of the variation of curvature of the tyre membrane

$$u_b = 2\pi\left(R - \left(h_t + \frac{h_b}{2}\right)\right) \qquad \textbf{8.142}$$

$$u_t = 2\pi\left(R - \frac{h_t}{2}\right) \qquad \textbf{8.143}$$

The energy loss $E_{loss,B}$ caused by hysteresis due to the variation of curvature of the sidewall, belt and tread when entering the contact area can be defined as (see Figure 8.16):

$$E_{loss,B} = \sum_i \left[\pi M_{i,B} \vartheta_{i,B} \, sin(\delta_i) \cdot \frac{u_i}{l_B}\right] \qquad i \in [s, b, t] \qquad \textbf{8.144}$$

where $M_{i,B}$ is the bending torque needed to rotate the sidewall, belt and tread by $\vartheta_{i,B}$ (sidewalls are counted twice) and l_B represents the length of the tyre circumference that is subjected to the torque. l_B can be set equal to half of the length l_c of the contact area. Only a part of the energy spent to bend the tyre is energy loss: this portion is approximated by $\sin(\delta_i)$. Instead, the elastic part of the bending torque is returned when exiting the contact path.

In order to quantify the bending torque $M_{i,B}$ the segment of belt involved is modelled as a straight bar. Bending torque $M_{i,B}$ needed to bend the tyre can then be derived by:

$$E_i I_{i,zy} \frac{\partial^2 v_i(x)}{\partial x^2} = M_{i,B} \qquad i \in [s, b, t] \qquad \textbf{8.145}$$

where v is the deflection of the bar:

$$v_i(x) = \frac{M_{i,B}}{E_i I_{i,zy}} \left(\frac{x^2}{2} + Ax + B \right) \qquad i \in [s, b, t] \qquad \textbf{8.146}$$

Imposing boundary conditions:

$$v_i(x)|_0 = 0 \qquad \textbf{8.147}$$

$$\left. \frac{\partial v_i(x)}{\partial x} \right|_0 = 0 \qquad i \in [s, b, t] \qquad \textbf{8.148}$$

the solution of the differential equation is:

$$v_i(x) = \frac{M_{i,B}}{E_i I_{i,zy}} \frac{x^2}{2} \qquad i \in [s, b, t] \qquad \textbf{8.149}$$

Subsequently, torques for sidewall, belt and tread at the extremum of the bar are ($x = l_b$ and $v(l_B) = z_b$ for the sidewall and $v(l_B) = z$ for the belt and the tread):

$$M_{s,B} = \frac{2z_b}{l_B^2} E_s I_{s,zy} \qquad \textbf{8.150}$$

$$M_{b,B} = \frac{2z}{l_B^2} E_b I_{b,zy} \qquad \textbf{8.151}$$

$$M_{t,B} = \frac{2z}{l_B^2} E_t I_{t,zy} \qquad \textbf{8.152}$$

where E_i are the elastic moduli and $I_{i,zy}$ the geometrical moments of inertia.

The rotations $\vartheta_{i,B}$ needed when bending the sidewall, the belt and the tread are respectively approximated by:

$$\vartheta_{s,B} = \tan^{-1}\left(\frac{(z_r + z_b)/2}{l_B}\right) \tag{8.153}$$

$$\vartheta_{b,B} = \tan^{-1}\left(\frac{(z_b + z)/2}{l_B}\right) \tag{8.154}$$

$$\vartheta_{t,B} = \tan^{-1}\left(\frac{(z_b + z)/2}{l_B}\right) \tag{8.155}$$

In Figure 8.17 the variation of the rolling resistance force (see equation 8.126) due to the elastic modulus of the sidewall is exemplarily shown for a 205/55 R16 run-flat tyre (see Table 8.1). For the chosen tyre a rolling resistance force of 30 N corresponds to an ISO rolling coefficient of 7.5 kg/t. On the one hand, higher values of elastic modulus reduce vertical deflection; on the other hand they slightly increase the bending torque. The dominance of one effect on the other changes with tyre load and inflation pressure.

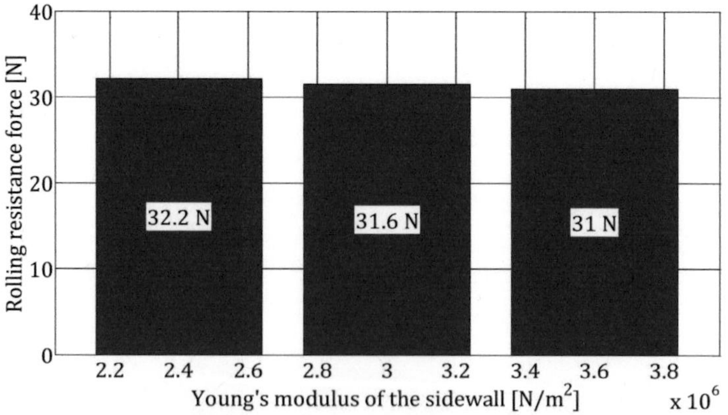

Figure 8.17: Variation of the rolling resistance force (physical tyre model)

9 Conflicts between the Requirements

In this section, the procedure for quantifying the conflicts between the requirements considering the physical constraints given by the tyre is presented. First, the choice of the design of experiment is motivated; then, the conflicts are identified and quantified using respectively correlation and sensitivity analyses.

9.1 Design of Experiment

The design of experiment (DoE) is chosen according to the type of analysis that should be performed. In our case the analysis concerns the identification and the quantification of conflicts between the requirements. For the identification of the conflicts correlation analyses are needed; for their quantification sensitivity analyses are used. Moreover, the results should be globally true, i.e. independent from the value of the design variables. Subsequently, the DoE should be generated varying all tyre design parameters (TDPs) at once. Of course, variations are within realistic physical ranges of tyres of passenger cars.

The design variables of the DoE are the 23 TDPs (form factors, reference ground pressure and factors q_1 to q_3 are held constant). For the first analyses all 23 parameters are chosen. Later considerations will lead to a reduction of the design variables to only 19 neglecting the four operating conditions p, F_z, v_x, $\mu_{a,ref}$. The experiments of the DoE are represented by different virtual tyres generated through the physical tyre model (see section 1). The more tyres are generated, the more precise are the results of the analysis.

Table 9.1: TDPs (exemplarily for 205/55 R16 run-flat tyre)

Name	Symbol	Unit	Range
Tyre radius	R	$[m]$	$[280\ 400] \cdot 10^{-3}$
Nominal width	w_n	$[m]$	$[195\ 325] \cdot 10^{-3}$
Sidewall height	h_s	$[m]$	$[80\ 130] \cdot 10^{-3}$
Sidewall thickness	w_s	$[m]$	$[7\ 15] \cdot 10^{-3}$
Belt height	h_b	$[m]$	$[5\ 9] \cdot 10^{-3}$
Tread height	h_t	$[m]$	$[4\ 7] \cdot 10^{-3}$
Void ratio	e_r	$[-]$	$[5\ 35] \cdot 10^{-2}$
Elastic modulus of sidewall	E_s	$[N/m^2]$	$[2.0\ 4.0] \cdot 10^6$
Elastic modulus of belt	E_b	$[N/m^2]$	$[1.5\ 3.5] \cdot 10^9$
Elastic modulus of tread	E_t	$[N/m^2]$	$[1.5\ 3.5] \cdot 10^6$
Rim stiffness	k_r	$[N/m]$	$[30\ 50] \cdot 10^6$
Air stiffness	k_a	$[N/m]$	$[50\ 90] \cdot 10^3$
Rim damping	d_r	$[Ns/m]$	[10 20]
Air damping	d_a	$[Ns/m]$	[0.09 0.15]
Sidewall damping	d_s	$[Ns/m]$	[30 40]
Belt damping	d_b	$[Ns/m]$	[15 25]
Tread damping	d_t	$[Ns/m]$	[25 35]
Rim mass	m_r	$[kg]$	[7 15]
Tyre mass	m	$[kg]$	[8 13]
Inflation pressure	p	$[Pa]$	$[2.0\ 3.5] \cdot 10^5$
Tyre load	F_z	$[N]$	$[3.0\ 5.0] \cdot 10^3$
Trajectory velocity	v_x	$[m/s]$	[60 100] / 3.6
Ref. adhesion coefficient	$\mu_{a,ref}$	$[-]$	[2.00 3.00]

The number of virtual tyres is raised until the indices of the correlation and sensitivity analysis converged to a stable value: for this DoE 5000 virtual tyres represent the best trade-off between time and precision. Above

5000 no significant improvement concerning the convergence of the indices is observed. The method for generating the DoE is chosen according to the homogeneity index (see section 3.4.2.4): the most homogeneous design having 23 design variables and 5000 experiments is generated by a latin hypercube. The variation ranges of each TDP are enlisted in Table 9.1.

Table 9.2: Selected set of functional tyre characteristics

Name	Symbol	Unit	Optimum
Rolling resistance force at reference conditions	$F_{r,0}$	$[N]$	↓
Cornering stiffness at zero slip angle	$K_{y,0}$	$[N/°]$	↑
Maximal lateral friction	$\mu_{y,max}$	$[N/N]$	↑
Position of maximal lateral friction	$\alpha_{max,\%}$	[%]	↑
Variation of cornering stiffness due to tyre load	$K'_{y,0,F_z}$	[%/N]	↓
Variation of maximal lateral friction due to tyre load	μ'_{y,max,F_z}	[%/N]	↓
Dynamic vertical stiffness	$K_{z,0}$	$[N/m]$	↓
Slope of dynamic vertical stiffness at 14 Hz	$K'_{z,14,f}$	$[Ns/m]$	↓
Average lateral force amplitudes at middle frequencies	$\bar{F}_{y,M}$	$[N]$	↓
Average vertical force amplitudes due to first eigenmode	$\bar{F}_{z,V}$	$[N]$	↓
Average vertical force amplitudes at middle frequencies	$\bar{F}_{z,M}$	$[N]$	↓
Average vertical force amplitudes due to air cavity	$\bar{F}_{z,C}$	$[N]$	↓

The design objective are the twelve FTCs derived in the previous section. They are enlisted in Table 9.2. They can be subdivided in four main categories: power loss (one FTC), lateral dynamics (five FTCs), ride comfort (two FTCs) and interior noise (four FTCs). It should be noticed, that for the analyses presented in this section the absolute value of FTC μ'_{y,max,F_z} is taken, in order to have all positive defined FTCs. Moreover, the FTC $\bar{F}_{z,M}$ is included in the analyses although its variation may be affected also by material and geometrical properties of the tyre that are not represented by the physical tyre model.

9.2 Analysis of the Tyre Design Parameter

In this section, the influence the tyre design parameters (TDP) have on the functional tyre characteristics (FTCs) is analysed. First, a global sensitivity analysis (GSA) concerning all 23 TDPs is presented; then, a GSA focusing only on 19 TDPs is performed. For the second analysis the four operating conditions p, F_z, v_x, $\mu_{a,ref}$ are neglected. Sobol's GSA is chosen, as it allows a global sensitivity evaluation (see section 3.5.2.1): the higher is Sobol's total index "T", the stronger is the influence of the TDPs on the FTCs. Strong influences are marked in green (Sobol's total index above 0.10). For the sake of completeness also Sobol's main index "M" is reported: the expert reader can use it to make further considerations. Moreover, the R^2-value of all artificial neuronal networks (ANN) needed to determine Sobol's indices are calculated: the results based on ANNs that have R^2-values below 0.95 are checked graphically.

9.2.1 DoE with 23 Design Variables

In this section, Sobol's GSA is based on the DoE concerning all 23 TDPs. The strength of the influence TDPs have on FTCs is presented in Table 9.3 and Table 9.4.

Table 9.3: GSA between TDPs and FTCs, 23 design variables (1/2)

	Power loss		Lateral dynamics									
	$F_{r,0}$ ($R^2=1$)		$K_{y,0}$ ($R^2=1$)		$\mu_{y,max}$ ($R^2=1$)		$\alpha_{max,\%}$ ($R^2=1$)		$K'_{y,0,F_z}$ ($R^2=0.99$)		μ'_{y,max,F_z} ($R^2=0.99$)	
	M	T	M	T	M	T	M	T	M	T	M	T
R	0.28	**0.34**	0	0	0	0	0	0	0	0	~0	~0
w_n	0.17	0.22	0.09	0.10	~0	~0	0.05	0.12	~0	~0	~0	~0
h_s	~0	~0	0.07	0.08	0	0	0.06	0.11	0	~0	0	0
w_s	~0	~0	0.20	**0.22**	0	0	0.22	**0.31**	~0	~0	0	0
h_b	0	0	0	0	0	0	0	0	0	0	0	0
h_t	~0	~0	0.10	0.11	~0	~0	0.11	0.14	~0	~0	0.05	0.06
e_r	0	0	~0	~0	~0	~0	0	~0	~0	~0	~0	~0
E_s	0	0	~0	0.05	0	0	0.06	0.08	~0	~0	~0	~0
E_b	0	0	0	0	0	0	0	0	0	0	0	0
E_t	~0	~0	0	~0	~0	~0	0	~0	~0	~0	0.13	0.13
p	0.05	0.06	0.11	0.13	~0	~0	0.07	0.07	~0	~0	0.09	0.10
F_z	0.18	0.22	0.20	**0.22**	~0	~0	0	~0	0.85	**0.85**	0.61	**0.62**
v_x	0.06	0.07	0	0	0	0	0	0	0	0	0	0
$\mu_{a,ref}$	0	0	0	0	0.88	**0.88**	0.09	0.14	0	0	0	0
k_r	0	0	0	0	0	0	0	0	0	0	0	0
k_a	0	0	0	0	0	0	0	0	0	0	0	0
d_r	0	0	0	0	0	0	0	0	0	0	0	0
d_a	0	0	0	0	0	0	0	0	0	0	0	0
d_s	0.05	0.06	0	0	0	0	0	0	0	0	0	0
d_b	0	0	0	0	0	0	0	0	0	0	0	0
d_t	0	0	0	0	0	0	0	0	0	0	0	0
m_r	0	0	0	0	0	0	0	0	0	0	0	0
m	0	0	0	0	0	0	0	0	0	0	0	0

Power loss (see Table 9.3) is characterized by tyre radius R, nominal width w_n and tyre load F_z. These TDPs influence the tyre deflection and the volume of the deformed mass, and, subsequently, the rolling resistance force.

Lateral dynamics is characterized by nominal width w_n, sidewall thickness w_s, sidewall height h_s, tread height h_t, inflation pressure p, tyre load F_z, elastic modulus of the tread E_t, and reference adhesion coefficient $\mu_{a,ref}$. These TDPs influence the lateral stiffness of the tyre and the contact area, and, subsequently, slip stiffness and maximal lateral friction.

Ride comfort (see Table 9.4) is characterized by tyre radius R, nominal width w_n, inflation pressure p, sidewall damping d_s and tyre mass m. Tyre radius, nominal width and inflation pressure affect mainly the absolute value of vertical dynamic stiffness at $0\ Hz$, whereas the sidewall damping and tyre mass influences the slope of the vertical dynamic stiffness by changing the position of the tyre's first vertical eigenfrequency.

Interior noise is characterized by tread height h_t, elastic modulus of the tread E_t, tyre load F_z, sidewall damping d_s and tyre mass m. Tread height, elastic modulus and tyre load of the tread influence the vertical stiffness of the tyre by varying its contact area and its stiffness. Sidewall damping changes the tyre's ability to absorb lateral oscillations. Tyre mass influence the vertical eigenfrequency and, subsequently, the response function to road excitation. It should be noticed, that tyre load and tyre mass are less sensitive then other parameters as, as already shown in experimental studies (see section 7.2.3).

The results of this first sensitivity analysis show that the operating conditions, especially tyre load, dominate several FTCs. This result is somehow trivial. Trivial is also the fact that reference adhesion coefficient characterizes the maximal transmissible lateral force. In order to focus on the influence of geometrical and material properties, in the section 9.2.2 the sensitivity analysis is repeated removing from the design variable set the TDPs concerning the operating conditions (p, F_z, v_x, $\mu_{a,ref}$).

Table 9.4: GSA between TDPs and FTCs, 23 design variables (2/2)

	Ride comfort				Interior noise							
	$K_{z,0}$ $(R^2 = 1)$		$K'_{z,14,f}$ $(R^2 = 1)$		$\bar{F}_{y,M}$ $(R^2 = 1)$		$\bar{F}_{z,V}$ $(R^2 = 0.98)$		$\bar{F}_{z,M}$ $(R^2 = 0.84)$		$\bar{F}_{z,C}$ $(R^2 = 0.96)$	
	M	T	M	T	M	T	M	T	M	T	M	T
R	0.11	0.11	~0	~0	0	0	~0	~0	~0	~0	~0	~0
w_n	0.62	**0.63**	~0	0.06	~0	~0	~0	~0	~0	0.08	~0	0.06
h_s	~0	~0	0	0	0	0	0	~0	~0	~0	0	~0
w_s	~0	~0	0	0	0	0	0	0	0	~0	0	0
h_b	0	0	0	0	0	0	0	0	0	~0	0	0
h_t	~0	~0	~0	~0	0.12	0.14	0.11	0.17	0.06	0.13	0.08	0.16
e_r	0	0	~0	~0	0.05	0.06	~0	0.08	~0	0.07	~0	0.07
E_s	~0	~0	0	0	0	0	0	0	0	~0	0	0
E_b	0	0	0	0	0	0	0	0	0	~0	0	0
E_t	~0	~0	0.08	0.09	0.25	**0.28**	0.26	**0.33**	0.13	**0.21**	0.15	**0.25**
p	0.15	0.15	0	~0	0.06	0.07	~0	0.07	~0	0.06	~0	0.08
F_z	~0	~0	~0	~0	0.09	0.11	0.10	0.14	0.05	0.11	0.06	0.12
v_x	0	0	0	0	0	0	0	0	0	~0	0	0
$\mu_{a,ref}$	0	0	0	0	0	0	0	0	0	~0	0	0
k_r	0	0	0	0	~0	0.05	~0	~0	0	~0	~0	~0
k_a	0	0	0	0	0	0	0	~0	~0	~0	~0	~0
d_r	0	0	0	0	0	0	0	0	0	~0	0	0
d_a	0	0	0	0	0	0	0	0	0	~0	0	0
d_s	0	0	0.63	**0.64**	0.10	0.11	~0	~0	~0	~0	~0	~0
d_b	0	0	0	0	0	0	0	0	0	~0	0	0
d_t	0	0	0	0	0	0	0	0	0	0	0	0
m_r	0	0	0	0	~0	~0	~0	~0	0	~0	0	~0
m	0	0	0.09	0.10	0.11	0.13	~0	0.10	0.06	0.12	0.08	0.15

9.2.2 DoE with 19 Design Variables

In this section, Sobol's GSA is based on the DoE with 19 TDPs. The strength of the influence TDPs have on FTCs is presented in Table 9.5 and Table 9.6.

Table 9.5: GSA between TDPs and FTCs, 19 design variables (1/2)

	Power loss		Lateral dynamics									
	$F_{r,0}$ ($R^2 = 1$)		$K_{y,0}$ ($R^2 = 1$)		$\mu_{y,max}$ ($R^2 = 1$)		$\alpha_{max,\%}$ ($R^2 = 1$)		$K'_{y,0,F_z}$ ($R^2 = 1$)		μ'_{y,max,F_z} ($R^2 = 1$)	
	M	T	M	T	M	T	M	T	M	T	M	T
R	0.49	**0.53**	0	0	~0	~0	-~0	0	~0	0.05	~0	~0
w_n	0.30	0.34	0.15	0.16	0.10	0.10	0.09	0.13	~0	0.09	0.09	0.11
h_s	~0	~0	0.13	0.14	0	0	0.07	0.14	~0	0.05	~0	~0
w_s	~0	~0	0.34	**0.37**	~0	0	0.23	**0.40**	0.13	0.16	~0	~0
h_b	0	0	0	0	0	0	0	0	0	0	0	0
h_t	~0	~0	0.17	0.18	0.07	0.08	0.13	0.18	0.16	0.16	0.20	0.20
e_r	0	0	~0	~0	0.62	**0.63**	~0	~0	0.06	0.07	0.10	0.10
E_s	~0	~0	0.08	0.09	0	0	0.08	0.11	0.07	0.08	~0	~0
E_b	0	0	0	0	0	0	0	0	0	0	0	0
E_t	~0	~0	~0	~0	0.17	0.18	-~0	~0	0.33	**0.34**	0.49	**0.49**
k_r	0	0	0	0	0	0	0	0	0	0	0	0
k_a	0	0	0	0	0	0	0	0	0	0	0	0
d_r	0	0	0	0	0	0	0	0	0	0	0	0
d_a	0	0	0	0	0	0	0	0	0	0	0	0
d_s	0.09	0.10	0	0	0	0	0	0	0	0	0	0
d_b	0	0	0	0	0	0	0	0	0	0	0	0
d_t	0	0	0	0	0	0	0	0	0	0	0	0
m_r	0	0	0	0	0	0	0	0	0	0	0	0
m	0	0	0	0	0	0	0	0	0	0	0	0

It should be remembered, that Sobol's indices are relative indices: the sum of the total Sobol indices "T" concerning one FTC is always equal to one. Subsequently, removing TDPs that have high sensitivity indices causes all other indices to rise.

Regarding the TDPs of power loss a new TDP is identified as relevant (see Table 9.5): the sidewall damping d_s, which influences directly the rolling resistance force.

Table 9.6: GSA between TDPs and FTCs, 19 design variables (2/2)

	Ride comfort				Interior noise							
	$K_{z,0}$ ($R^2 = 1$)		$K'_{z,14,f}$ ($R^2 = 1$)		$\bar{F}_{y,M}$ ($R^2 = 1$)		$\bar{F}_{z,V}$ ($R^2 = 0.99$)		$\bar{F}_{z,M}$ ($R^2 = 0.94$)		$\bar{F}_{z,C}$ ($R^2 = 0.98$)	
	M	T	M	T	M	T	M	T	M	T	M	T
R	0.14	0.14	~0	~0	0	0	0	~0	~0	~0	~0	~0
w_n	0.73	**0.73**	0.06	0.07	~0	~0	~0	~0	~0	0.09	~0	0.07
h_s	~0	~0	0	0	0	0	~0	~0	~0	~0	~0	~0
w_s	~0	~0	0	0	0	0	0	0	0	0	0	0
h_b	0	0	0	0	0	0	0	0	0	~0	0	0
h_t	0	~0	~0	~0	0.16	0.18	0.11	0.21	0.08	0.17	0.13	0.20
e_r	0	0	~0	~0	0.07	0.08	0.06	0.09	~0	0.09	0.05	0.09
E_s	0.05	0.05	0	0	0	0	0	0	0	0	0	0
E_b	0	0	0	0	0	0	0	0	0	0	0	0
E_t	~0	~0	0.07	0.07	0.33	**0.35**	0.32	**0.42**	0.16	**0.29**	0.20	**0.32**
k_r	0	0	0	0	~0	0.05	~0	~0	0	~0	~0	~0
k_a	0	0	0	0	0	0	~0	~0	~0	~0	~0	~0
d_r	0	0	0	0	0	0	0	0	0	0	0	0
d_a	0	0	0	0	0	0	0	0	0	0	0	0
d_s	0	0	0.71	**0.73**	0.12	0.13	~0	~0	~0	~0	~0	~0
d_b	0	0	0	0	0	0	0	0	0	0	0	0
d_t	0	0	0	0	0	0	0	0	0	0	0	0
m_r	0	0	0	0	~0	~0	~0	~0	0	~0	~0	~0
m	0	0	0.06	0.06	0.13	0.15	~0	0.12	0.08	0.18	0.11	0.18

Concerning lateral dynamics, beside nominal width w_n, sidewall thickness w_s, sidewall height h_s, tread height h_t, also void ratio e_r, elastic modulus of the tread E_t and elastic modulus of the sidewall E_s gain importance.

Concerning ride comfort and interior noise (see Table 9.6), the relevant TDPs match with those identified in the previous analysis.

This second GSA focuses better on the influence of TDPs concerning geometrical and material properties, because the four TDPs (p, F_z, v_x, $\mu_{a,ref}$) describing the operating conditions are excluded from the analysis, i.e.

held constant. By doing so, the ranking of TDPs remains the same, but their importance is highlighted.

9.3 Analysis of Conflicts

In this section, first, a correlation analysis is used to identify if two functional tyre characteristics (FTCs) are correlated positively or negatively; then a sensitivity analysis quantifies the strength of their relation.

9.3.1 Identification of Conflicts

To identify the conflicts, a correlation analysis is needed. As the correlation analysis serves only to determine whether the FTCs are correlated positively or negatively, it is sufficient to rank the values of the FTCs. Therefore, Spearman's correlation analysis is chosen (see section 3.5.1.2).

The conflicts are identified by a negative correlation index and marked in red if Spearman's index is below -0.10. Conflicts are always characterized by negative correlation, because, the sign of the FTCs, which are all defined positive, is adjusted according to their optima (see Table 9.2) so that less is better. To check the results, the p-value is calculated: if below 0.05, the results are statistically significant (see section 3.5.1). The results based on correlation analysis having p-values above 0.05 are checked graphically.

9.3.1.1 DoE with 23 Design Variables

In this section, Spearman's correlation analysis is based on the DoE concerning all 23 TDPs. The results are presented in Table 9.7 and Table 9.8. All requirements are in conflict with each other. Especially power loss with ride comfort and lateral dynamics with interior noise (see Table 9.7 and Table 9.8). Of course, also FTCs of one requirement are in conflict with each other, e.g. $\mu_{y,max}$ and $\alpha_{max,\%}$.

Table 9.7: Identification of conflicts, 23 design variables (1/2)

		Power loss	Lateral dynamics				
		$F_{r,0}$	$K_{y,0}$	$\mu_{y,max}$	$\alpha_{max,\%}$	$K'_{y,0,F_z}$	μ'_{y,max,F_z}
Power loss	$F_{r,0}$						
Lateral dynamics	K_y	-0.12					
	$\mu_{y,max}$	~0	~0				
	$\alpha_{max,\%}$	0.07	0.83	-0.32			
	$K'_{y,0,F_z}$	-0.32	0.44	-0.08	0.09		
	μ'_{y,max,F_z}	-0.57	0.51	-0.05	0.20	0.54	
Ride comfort	$K_{z,0}$	-0.62	-0.35	~0	-0.35	-0.25	0.15
	$K'_{z,14,f}$	0.25	0.27	~0	0.23	0.09	0.28
Interior noise	$\bar{F}_{y,M}$	0.21	-0.44	~0	-0.33	-0.10	-0.64
	$\bar{F}_{z,V}$	0.11	-0.51	~0	-0.39	-0.13	-0.66
	$\bar{F}_{z,M}$	0.07	-0.40	~0	-0.30	-0.10	-0.52
	$\bar{F}_{z,C}$	~0	-0.45	~0	-0.34	-0.12	-0.58

Table 9.8: Identification of conflicts, 23 design variables (2/2)

		Ride comfort		Interior noise			
		$K_{z,0}$	$K'_{z,14,f}$	$\bar{F}_{y,M}$	$\bar{F}_{z,V}$	$\bar{F}_{z,M}$	$\bar{F}_{z,C}$
Ride comfort	$K_{z,0}$						
	$K'_{z,14,f}$	-0.24					
Interior noise	$\bar{F}_{y,M}$	0.11	-0.14				
	$\bar{F}_{z,V}$	0.23	-0.26	0.86			
	$\bar{F}_{z,M}$	0.24	-0.28	0.88	0.85		
	$\bar{F}_{z,C}$	0.27	-0.21	0.93	0.84	0.93	

9.3.1.2 DoE with 19 Design Variables

In this section, Spearman's correlation analysis is based on the DoE with 19 TDPs. The results are presented in Table 9.9 and Table 9.10.

The conflicts identified by the correlation analysis involving 19 TDPS are very similar to the ones identified by the analysis involving all 23 TDPs. The major changes regard $F_{r,0}$, which is now in conflict also with FTC of interior noise, and $K'_{y,0,F_z}$, which is no more in conflict with interior noise. This should not surprise: the choice of different sets of TDPs leads to two different results affecting both the strength and the presence of a conflict. Concerning $K'_{y,0,F_z}$, the GSA presented in section 9.2.1 shows that $K'_{y,0,F_z}$ and all FTCs of interior noise are dependent from tyre load. By excluding tyre load from the analysis, also the conflict vanishes.

Table 9.9: Identification of conflicts, 19 design variables (1/2)

		Power loss	Lateral dynamics				
		$F_{r,0}$	$K_{y,0}$	$\mu_{y,max}$	$\alpha_{max,\%}$	$K'_{y,0,F_z}$	μ'_{y,max,F_z}
Power loss	$F_{r,0}$						
Lateral dynamics	K_y	0.30					
	$\mu_{y,max}$	0.20	~0				
	$\alpha_{max,\%}$	0.27	0.99	-0.06			
	$K'_{y,0,F_z}$	0.28	0.18	0.16	0.16		
	μ'_{y,max,F_z}	-0.28	~0	-0.22	0.07	-0.94	
Ride comfort	$K_{z,0}$	-0.80	-0.59	-0.23	-0.56	-0.30	0.26
	$K'_{z,14,f}$	0.47	0.24	~0	0.23	-0.16	0.20
Interior noise	$\bar{F}_{y,M}$	~0	-0.28	0.09	-0.29	0.61	-0.68
	$\bar{F}_{z,V}$	-0.07	-0.34	0.11	-0.35	0.62	-0.71
	$\bar{F}_{z,M}$	-0.09	-0.29	0.06	-0.30	0.50	-0.56
	$\bar{F}_{z,C}$	-0.22	-0.30	~0	-0.30	0.56	-0.60

Table 9.10: Identification of conflicts, 19 design variables (2/2)

		Ride comfort		Interior noise			
		$K_{z,0}$	$K'_{z,14,f}$	$\bar{F}_{y,M}$	$\bar{F}_{z,V}$	$\bar{F}_{z,M}$	$\bar{F}_{z,C}$
Ride comfort	$K_{z,0}$						
	$K'_{z,14,f}$	-0.28					
Interior noise	$\bar{F}_{y,M}$	0.18	-0.05				
	$\bar{F}_{z,V}$	0.22	-0.13	0.83			
	$\bar{F}_{z,M}$	0.24	-0.18	0.87	0.83		
	$\bar{F}_{z,C}$	0.35	-0.11	0.92	0.79	0.93	

9.3.2 Quantification of Conflicts

To quantify the conflicts a global sensitivity analysis (GSA) is needed. Sobol's GSA is chosen (see section 3.5.2.1): the higher is Sobol's total index "T", the stronger is the relation between two FTCs. Only the conflicts identified previously by the correlation analysis are analysed. Strong influences are marked in green (Sobol's total index above 0.10). Moreover, the R^2-value of all artificial neuronal networks (ANN) needed to determine Sobol's indices are calculated: due to the complexity of the relation between the FTCs, several results are based on ANNs that have R^2-values below 0.95. Subsequently, the numerical values are checked graphically.

It should be underlined, that Sobol's indices are relative indices and calculated between all inputs (i.e. all selected FTCs enlisted in the first column of the GSA tables) and respectively one output (i.e. the FTC enlisted in the first row of the GSA tables). Subsequently, it is not possible to compare the strength of the conflicts of different columns: only the ranking of the conflicts quantified in different columns can be compared.

9.3.2.1 DoE with 23 Design Variables

In this section, Sobol's GSA is based on the DoE concerning all 23 TDPs. The strength of the conflicts is presented in Table 9.11 and Table 9.12.

All requirements are in conflict mainly due to R, w_n, h_t, E_t and F_z (see Table 9.11 and Table 9.3 for power loss and lateral dynamics, and Table 9.12 and Table 9.4 for ride comfort and interior noise). Minor conflicts are generated due to w_s, h_s, p, d_s and m.

Concerning power loss (see Table 9.11), one of the strongest conflict exists between $F_{r,0}$ and $K_{z,0}$: it is mainly caused by R and w_n. Big and large tyres reduce vertical deflections, but increase vertical stiffness, generating poor ride comfort. A minor conflict is given by inflation pressure.

Concerning lateral dynamics (see Table 9.11), one of the strongest conflict exists between $K_{y,0}$ and $K_{z,0}$: it is mainly caused by w_n.

Table 9.11: Quantification of conflicts, 23 design variables (1/2)

	$F_{r,0}$ ($R^2 = 0.72$)		$K_{y,0}$ ($R^2 = 0.51$)		$\mu_{y,max}$ ($R^2 = 0.16$)		$\alpha_{max,\%}$ ($R^2 = 0.23$)		$K'_{y,0,F_z}$ ($R^2 = 0.59$)		μ'_{y,max,F_z} ($R^2 = 0.82$)	
	M	T	M	T	M	T	M	T	M	T	M	T
$F_{r,0}$			0	0					0.25	0.25	0.34	**0.37**
$K_{y,0}$	~0	0.08										
$\mu_{y,max}$							0.29	**0.30**	0	0	0	0
$\alpha_{max,\%}$					0.79	**0.82**						
$K'_{y,0,F_z}$	0.08	0.16			~0	0.06						
μ'_{y,max,F_z}	0.06	0.13			~0	0.12						
$K_{z,0}$	0.51	**0.64**	0.29	**0.31**			0.24	0.25	0.35	**0.35**		
$K'_{z,14,f}$												
$\bar{F}_{y,M}$			0.19	0.20			0.14	0.14	0.10	0.17	~0	0.22
$\bar{F}_{z,V}$			0.22	0.25			0.15	0.16	~0	0.10	~0	0.23
$\bar{F}_{z,M}$			0.05	0.11			0.05	0.07	0	0	~0	0.05
$\bar{F}_{z,C}$			~0	0.08			~0	0.08	~0	0.10	0.05	0.13

Large tyres increase slip stiffness, but also vertical stiffness, generating poor ride comfort. Further conflicts exist between lateral dynamics and interior noise: they are caused by h_t, E_t and F_z.

Concerning ride comfort (see Table 9.12), a conflict exists between $K_{z,0}$ and $F_{r,0}$, as already mentioned before. A second interesting conflict is generated between $K'_{z,14,f}$ and $\bar{F}_{z,M}$: it is caused by m. Low tyre mass causes the tyre's first vertical eigenfrequency to shift to higher frequencies (and, subsequently, to reduce slope $K'_{z,14,f}$), but increase the magnitude of the transmitted vertical forces in the middle frequency range, generating poor interior noise.

Concerning interior noise (see Table 9.12), one significant conflict exists with μ'_{y,max,F_z}: it is generated by E_t and F_z. Low tread stiffness and low tyre load improve interior noise, but increase the influence tyre load has on lateral grip, generating poor lateral dynamics. On the contrary, conflicts with power loss and with ride comfort are of minor importance.

Table 9.12: Quantification of conflicts, 23 design variables (2/2)

	$K_{z,0}$ ($R^2 = 0.73$)		$K'_{z,14,f}$ ($R^2 = 0.57$)		$\bar{F}_{y,M}$ ($R^2 = 0.57$)		$\bar{F}_{z,V}$ ($R^2 = 0.64$)		$\bar{F}_{z,M}$ ($R^2 = 0.43$)		$\bar{F}_{z,C}$ ($R^2 = 0.51$)	
	M	T	M	T	M	T	M	T	M	T	M	T
$F_{r,0}$	0.32	**0.32**										
$K_{y,0}$	0.07	0.22			0.19	0.23	0.20	0.29	0.09	0.19	0.17	0.23
$\mu_{y,max}$												
$\alpha_{max,\%}$	0.16	0.26			0.05	0.16	0.05	0.16	0.09	0.19	0.07	0.17
$K'_{y,0,F_z}$	~0	0.17			0.05	0.17	0.15	0.17	0.07	0.18	~0	0.10
μ'_{y,max,F_z}					0.43	**0.44**	0.33	**0.33**	0.35	**0.38**	0.39	**0.45**
$K_{z,0}$			0.08	0.09								
$K'_{z,14,f}$	~0	~0			0	0	~0	0.05	~0	0.06	~0	0.05
$\bar{F}_{y,M}$			0.17	0.23								
$\bar{F}_{z,V}$			0.14	0.17								
$\bar{F}_{z,M}$			0.23	**0.28**								
$\bar{F}_{z,C}$			0.06	0.14								

9.3.2.2 DoE with 19 Design Variables

In this section, Sobol's GSA is based on the DoE with 19 TDPs. The strength of the conflicts is presented in Table 9.13 and Table 9.14. It should be noted, that the FTC $\alpha_{max,\%}$ is exclude from the GSA as it is strongly correlated to $K_{y,0}$ (see Table 9.9): their conflicts would be the same.

The following TDPs are those generating the main conflicts between the requirements: R, w_n, h_t and E_t (see Table 9.13 and Table 9.5 for power loss and lateral dynamics, Table 9.14 and Table 9.6 for ride comfort and interior noise). Minor conflicts are generated due to w_s, h_s, d_s, and m. This results are comparable to those of the analysis concerning all 23 TDPs.

Concerning power loss (see Table 9.13), one of the strongest conflict exists between $F_{r,0}$ and $K_{z,0}$: it is caused by R and w_n. Big and large tyres reduce vertical deflections, but increase vertical stiffness, generating poor ride comfort. Similar considerations are true for the minor conflict between $F_{r,0}$ and $\bar{F}_{z,M}$.

Table 9.13: Quantification of conflicts, 19 design variables (1/2)

	$F_{r,0}$ ($R^2 = 0.69$)		$K_{y,0}$ ($R^2 = 0.54$)		$\mu_{y,max}$ ($R^2 = 0.15$)		$\alpha_{max,\%}$		$K'_{y,0,F_z}$ ($R^2 = 0.91$)		μ'_{y,max,F_z} ($R^2 = 0.96$)	
	M	T	M	T	M	T	M	T	M	T	M	T
$F_{r,0}$							correlated to $K_{y,0}$				$\sim$0	0.09
$K_{y,0}$												
$\mu_{y,max}$											$\sim$0	0.11
$\alpha_{max,\%}$												
$K'_{y,0,F_z}$											0.39	**0.44**
μ'_{y,max,F_z}	$\sim$0	0.09			0.32	0.40			0.98	**0.98**		
$K_{z,0}$	0.59	**0.59**	0.29	**0.32**	0.54	**0.60**			0	0		
$K'_{z,14,f}$									0	0		
$\bar{F}_{y,M}$			0.07	0.17							0.06	0.10
$\bar{F}_{z,V}$	0.12	0.13	0.09	0.20							0.09	0.12
$\bar{F}_{z,M}$	0	0	0.17	0.17							0.05	0.12
$\bar{F}_{z,C}$	$\sim$0	0.17	0.06	0.14							$\sim$0	0.10

Concerning lateral dynamics (see Table 9.13), conflicts exist between $K_{y,0}$ and $K_{z,0}$ as well as $\mu_{y,max}$ and $K_{z,0}$: both are mainly caused by w_n. Large tyres increase lateral stiffness and lateral grip, but also vertical stiffness, generating poor ride comfort. Further conflicts exist between lateral dynamics and interior noise, especially concerning $K_{y,0}$ and μ'_{y,max,F_z}: they are caused by h_t and E_t.

Concerning ride comfort (see Table 9.14), one conflict exists between $K_{z,0}$ and $F_{r,0}$, as already mentioned before. Further conflicts involve $K'_{z,14,f}$ and interior noise FTCs, e.g. $\bar{F}_{y,M}$ and $\bar{F}_{z,M}$; they are caused by m and d_s.

Low tyre mass and high damping causes the tyre's first vertical eigenfrequency to enlarge and to shift to higher frequencies (and, subsequently, to reduce slope $K'_{z,14,f}$), but increase the magnitude of the transmitted vertical forces in the middle frequency range, generating poor interior noise.

Table 9.14: Quantification of conflicts, 19 design variables (2/2)

	$K_{z,0}$ ($R^2 = 0.85$)		$K'_{z,14,f}$ ($R^2 = 0.62$)		$\bar{F}_{y,M}$ ($R^2 = 0.59$)		$\bar{F}_{z,V}$ ($R^2 = 0.73$)		$\bar{F}_{z,M}$ ($R^2 = 0.51$)		$\bar{F}_{z,C}$ ($R^2 = 0.63$)	
	M	T	M	T	M	T	M	T	M	T	M	T
$F_{r,0}$	0.58	**0.62**					~0	0.07	0.05	0.09	0.08	0.13
$K_{y,0}$	0.17	0.22			0.22	0.24	0.10	0.21	0.09	0.10	0.10	0.14
$\mu_{y,max}$	0	0										
$\alpha_{max,\%}$												
$K'_{y,0,F_z}$	0.07	0.12	~0	0.08								
μ'_{y,max,F_z}					0.71	**0.74**	0.52	**0.63**	0.72	**0.73**	0.56	**0.67**
$K_{z,0}$			0.08	0.15								
$K'_{z,14,f}$	~0	~0			0	0	~0	0.09	~0	0.09	~0	0.06
$\bar{F}_{y,M}$			0.10	0.17								
$\bar{F}_{z,V}$			~0	0.08								
$\bar{F}_{z,M}$			0.16	**0.23**								
$\bar{F}_{z,C}$			0.12	0.15								

Concerning interior noise (see Table 9.14), the significant conflicts exist between FTCs of interior noise and μ'_{y,max,F_z}: it is generated by E_t and h_t. Low tread stiffness and high tread height improve interior noise, but increase the influence tyre load has on lateral grip, generating poor lateral dynamics. Same effects are observed for the conflict between all FTCs of interior noise and $K_{y,0}$. On the contrary, conflicts with power loss and with ride comfort are of minor importance.

9.4 Case Study: Pareto-Optimal Set

In this section, the Pareto-optimal set concerning two FTCs should be exemplarily shown. The Pareto-optimal set defines the border of the region of all possible design solution. The optimal solutions are those and only those lying on this border (Mastinu, Gobbi, & Miano, 2004): these solutions represent the best compromise in optimising (minimising or maximising) two FTCs. All solutions of the Pareto-optimal set are equally optimal: only by fixing weights for the FTCs it would be possible to choose one design solution of the Pareto-optimal set. A solution $\boldsymbol{x}_i$ is Pareto-optimal, if is satisfies the following conditions:

$$\nexists \boldsymbol{x}_j\colon \begin{cases} f_k(\boldsymbol{x}_j) \leq f_k(\boldsymbol{x}_i) & \forall \boldsymbol{k} = \mathbf{1} \ldots \boldsymbol{n_{OF}} \\ \exists l\colon f_l(\boldsymbol{x}_j) < f_l(\boldsymbol{x}_i) & \boldsymbol{k} \in [\mathbf{1} \ldots \boldsymbol{n_{OF}}] \end{cases} \qquad \mathbf{9.1}$$

where n_{OF} is the number of objective functions (in our case the FTCs that have to be minimised).

The Pareto-optimal set can be calculated using e.g. a genetic algorithm and implementing a multi-objective optimisation. Alternatively, it could be graphically represented by a curve delimiting the region of a scatter plot. Moreover, a Pareto-optimal set has a Pareto-optimal curve that often needs an optimisation to be determined. In this section, only an example of the Pareto-optimal set should be shown.

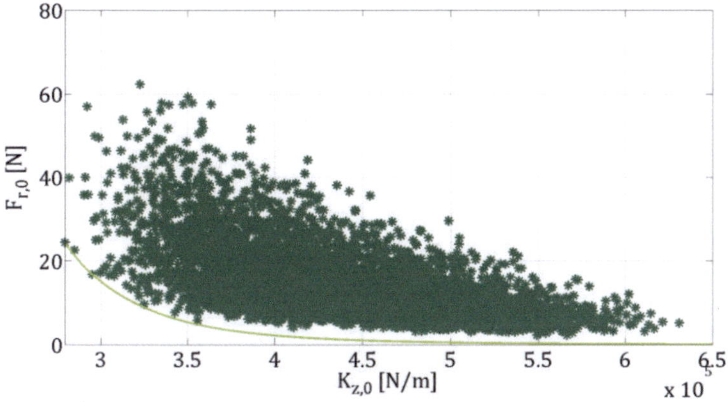

Figure 9.15: Pareto-optimal set depicting the conflict between power loss and ride comfort

In Figure 9.15 the conflict between rolling resistance force $F_{r,0}$ and vertical dynamic stiffness $K_{z,0}$ is represented by 5000 different tyres generated through the physical tyre model. The Pareto-optimal curve is approximate by a hyperbole. A hyperbole is chosen as both FTCs have to be minimised: in fact, the lower the rolling resistance force and the lower the dynamic vertical stiffness the more requirements concerning power loss and ride comfort are satisfied. In order to determine the Pareto-optimal curve the slope and the exponent of the generic hyperbole y_P:

$$y_P(x) = \frac{a}{x^b} \quad \mathbf{9.2}$$

are set to make the hyperbole intersect two extrema of the Pareto-optimal set:

$$y_P(2{,}79 \cdot 10^5) = 24.4$$
$$y_P(2{,}95 \cdot 10^5) = 16.8 \quad \mathbf{9.3}$$

Then, the hyperbole is:

$$y_P(x) = \frac{6.8 \cdot 10^{37}}{x^{6.69}} \quad \mathbf{9.4}$$

A graphical representation of the Pareto-optimal set is a very useful instrument to show conflicts and to quantify how far an existing tyre may be from a possible optimal solution.

9.5 Summary

In section 9 the conflicts between the requirements concerning power loss, lateral dynamics, ride comfort and interior noise are identified and quantified. Two different design variable sets are chosen: the first consists of all 23 tyre design parameters (TDPs), the second of only 19 TDPs. The number of experiments is 5000 and the generation method that guarantees the most homogeneous design of experiments (DoEs) for both designs is the latin hypercube.

After having generated the DoEs, the influence each TDP has on the FTCs is analysed through a sensitivity analysis. Then, the conflicts are identified by Spearman's correlation analysis and quantified by Sobol's sensitivity analysis. The statistical significance of the correlation indices is determined by the p-value; the quality of the artificial neuronal network used to calculate the sensitivity indices is determined by the R^2-value.

The results of the analysis of conflicts are schematically summarised in Table 9.11, Table 9.12, Table 9.13 and Table 9.14. Power loss is in conflict with all requirements. The strongest conflicts occur between the other three requirements: lateral dynamics, ride comfort and interior noise.

The operating conditions have a strong influence on the conflicts, especially inflation pressure p and tyre load F_z. Concerning the pure geometrical and material properties of the tyre, conflicts are generated mainly due to tyre radius R, nominal width w_n, tread height h_t and elastic modulus of the tread E_t. Minor conflicts are generated due to w_s, h_s, d_s and m.

Power loss is in conflict with FTCs of ride comfort ($F_{r,0}$, $K_{z,0}$) and interior noise ($\bar{F}_{z,V}$). These conflicts are caused by R and w_n. Big and large tyres reduce vertical deflections, but increase vertical stiffness, generating poor ride comfort.

Lateral dynamics is in conflict with FTCs of ride comfort ($K_{z,0}$) and interior noise ($\bar{F}_{y,M}$, $\bar{F}_{z,V}$, $\bar{F}_{z,M}$, $\bar{F}_{z,C}$). These conflicts are caused by w_n, h_t, and E_t. Larger tyres and stiffer treads increase lateral stiffness and lateral grip, but increase vertical stiffness, generating poor ride comfort and interior noise.

Ride comfort is in conflict with FTCs of power loss ($F_{r,0}$) as well as of lateral dynamics (K_y) and interior noise ($\bar{F}_{y,M}$, $\bar{F}_{z,V}$). The conflicts are generated mainly by R, w_n, d_s and m. Slim and small tyres reduce vertical stiffness, generating good ride comfort, but increase lateral and vertical deformation generating poor power loss and lateral dynamics. High damping and low tyre mass causes the tyre's first vertical eigenfrequency to shift to higher frequencies (and, subsequently, to reduce slope $K'_{z,14,f}$), but increase the magnitude of the transmitted vertical forces in the middle frequency range, generating poor interior noise.

Interior noise is in conflict with FTCs of lateral dynamics (K_y, μ'_{y,max,F_z}): it is generated by h_t and E_t. High tread height and low tread stiffness improve interior noise, but increase the influence tyre load has on lateral grip, generating poor lateral dynamics. Same effects are observed for the conflict between all FTCs of interior noise and K_y. On the contrary, conflicts with power loss and with ride comfort are less strong.

10 Conclusions and Outlook

The present thesis focuses on the tyre, especially on its influence on four requirements of the vehicle development, namely power loss, lateral dynamics, ride comfort and interior noise. The objective of the thesis is the quantification of conflicts between four selected requirements considering the physical constraints given by the tyre. In this last section the conclusions and the added scientific value are pointed out; then, an outlook on further development is given.

After a brief introduction (see section 1), the state of the art (see section 1) and the choice of methods (see section 1) is presented. In section 3.1 the definition of "tyre design parameter" (TDP), "functional tyre characteristic" (FTC), "objective manoeuvre criterion" (OMC) and "subjective manoeuvre index" (SMI) is given. The requirements power loss, lateral dynamics, ride comfort and interior noise are analysed separately in sections 1 to 1. For each requirement a set of OMC is defined and manoeuvres to evaluate its OMC are selected. Then, FTCs needed to describe the influence tyres have on the OMC are defined and the correlations between FTCs and OMC are analysed through graphical and numerical sensitivity analyses.

Concerning power loss (see section 1), first, a physical definition is introduced and a formulation for an outer drum test bench is derived. Then, a new parameterization method for Magic Formula using the Tyre Fitter and an ad-hoc measurement routine is described. The manoeuvre used to evaluate the average customer power loss is presented in section 4.2.1. The power loss is defined as OMC and the rolling resistance force $F_{r,0}$ is chosen as most significant FTC; they are described by equation 4.13 and in Table 4.13.

Concerning lateral dynamics (see section 1), the choice of manoeuvres and of OMC necessary to evaluate vehicle reaction are presented in section

5.1. The "Vehicle Dynamics Formulae" (VDF) are introduced to define the set of possible OMC. Then, according to an objectivation study at the dynamic driving simulator the OMC necessary to reproduce the lateral dynamics are identified. The 15 most significant OMC and the five relevant FTCs for lateral dynamics (the main FTC is the cornering stiffness $K_{y,0}$) are enlisted in Table 5.27 and Table 5.32.

Concerning ride comfort (see section 1), the choice of manoeuvres and of OMC needed to evaluate vehicle bouncing and pitching as well as the wheel-suspension system vertical accelerations are presented in section 6.1. Then, according to a virtual design of experiment the most important FTCs are identified. Twelve OMC and two relevant FTCs are chosen as most important for ride comfort (the two FTC are the dynamic vertical stiffness $K_{z,0}$ and its slope $K'_{z,14,f}$); they are enlisted in Table 6.6 and Table 6.13.

Concerning interior noise (see section 1), the choice of manoeuvres and of OMC used to evaluate low frequency (hum, low rumble, drone) and high frequency phenomena (high rumble, air cavity) as well as the tyre spectrum at the wheel hub are presented in section 7.1. Then, according to an experimental design based on 96 measurements at the acoustic tyre test bench and on a test track the main FTCs are identified. The four OMC and the five relevant FTCs for interior noise are enlisted in Table 7.4 and Table 7.24.

In order to quantify the conflicts between the requirements a physical tyre model is developed (see section 1). It is composed of three main sub-models: a three mass model used to describe ride comfort and interior noise (in vertical and lateral direction), a brush model used to describe the quasi-static lateral dynamics and a hysteresis model. The physical tyre model reproduces the characteristics of power loss, lateral dynamics, ride comfort and interior noise for different combinations of TDPs and allows to evaluate the related FTCs as defined in sections 1 to 1. Using the physical tyre model it is possible to identify and quantify the conflicts between the requirements (see section 9). Moreover, knowing which TDP influences the FTCs most and which FTCs are in conflict, it is possible to derive which TDPs generate the conflicts. The results of the analyses are summarised in Table 9.11, Table 9.12, Table 9.13 and Table 9.14. In general, all requirements are in conflict with each other, especially lateral dynamics with ride comfort and interior noise. On the contrary, ride comfort and interior noise are mainly correlated.

The method proposed in the present thesis is composed of FTCs, a physical tyre model and a procedure for identifying and quantifying their conflicts. The FTCs are objective quantities that can be derived from any tyre model simulation or tyre measurement; they are a "common language" for communicating tyre characteristics. The physical tyre model and the proposed procedure allow to evaluate and visualise conflicts (see section 1 and 9) as a function of geometrical and material properties of the tyre (TDPs). The generated know-how supports decision-making in the pre-development phase contributing in reducing time and costs of the tyre and vehicle development.

Further developments should focus on the quantification of the robustness of the method and enlarge the requirement portfolio. The implementation of algorithms to calculate solution spaces for the TDPs improves the visualisation and the evaluation of the conflicts. Moreover, adding requirements (e.g. longitudinal dynamics, wet performance and tyre abrasive wear) and combing the proposed FTCs with those of other subcomponents (e.g. wheel-suspension system) increases the potential of the method guaranteeing a more comprehensive view of the complex tyre development process and allowing OEMs to better meet customer needs keeping pace with the fast growing market and the increasing competition.

Bibliography

Aboutorabi, H., & Kung, L. (2012). Application of Coupled Structural Acoustic analysis and Sensitivity Calculation to a Tire Noise problem. *Tire Science and Technology, 40*(1), 25-41.

Ammon, D., Gnadler, R., Mäckle, G., & Unrau, H.-J. (2004). Ermittlung der Reibwerte von Gummistollen zur genauen Parametrierung von Reifenmodellen. *ATZ*, 694-701.

Anfosso-Lédée, F., & Pichaud, Y. (2007). Temperature effect on tyre–road noise. *Applied Acoustics, 68*, 1–16.

Archer, G., Saltelli, A., & Sobol, I. (1997). Sensitivity measures, ANOVA like techniques and the use of bootstrap. *Journal of Statistical Computation and Simulation*(58), 99-120.

Ascough II, J., Green, T., Ma, L., & Ahjua, L. (2005). Key Criteria and Selection of Sensitivity Analysis Methods Applied to Natural Resource Models. *Modsim 2005 International Congress on Modeling and Simulation, Modeling and Simulation Society of Australia and New Zealand*, (S. 2463–2469). Melbourne.

Auto Bild. (2009). *Auto Bild*(10), 66-70.

Bahnert, T., Lienkamp, M., & Vogel, F. (2012). Experimental and numerical characterisation of tyres in terms of rolling noise. *3rd International Munich Chassis Symposium* (S. 807-824). Munich: Springer Verlag.

Bakker, E., Nyborg, L., & Pacejka, H. (1987). Tire Modelling for Use in Vehicle Dynamics Studies. *SAE Technical Paper 870421*, 1-15.

Bakker, E., Pacejka, H., & Lidner, L. (1989). A New Tire Model with an Application in Vehicle Dynamics Studies. *SAE Technical Paper 890087.*

Bayle, P., Forissier, J., & Lafon, S. (1993). A New Tire Model for Vehicle Dynamics Simulation. *Automotive Technology International,* 193–198.

Bergman, W. (1973). Measurement and Subjective Evaluation of Vehicle Handling. *SAE Technical Paper 730492.*

Bernard, J., Segel, L., & Wild, R. (1977). Tire shear force generation during combined steering and braking maneuvres. *SAE Technical Paper 770852.*

Blundell, M., & Harty, D. (2007). Intermediate tyre model for vehicle handling simulation. *Journal of Multi-doby Dynamics, 221,* 41-62.

Bobbert, G. (1988). Schwingungsauswirkung auf den Menschen im fahrenden Fahrzeug. In W. Stühler, *Fahrzeug Dynamik* (S. 135-151). Braunschweig: Vieweg+Teubner Verlag.

Botev, S. (2008). *Digitale Gesamtfahrzeugabstimmung für Ride und Handling.* Düsseldorf: VDI Verlag.

Bradley, C., & Delaval, A. (2013). On-Road Fuel Consumption Testing to Determine the Sensitivity Coefficient Relating Changes in Fuel Consumption to Changes in Tire Rolling Resistance. *Tire Science and Technology, 41*(1), 2-20.

Brandstätter, M. (2013). Tieffrequente Geräusche in einem Kraftfahrzeug bei Unebenheitsanregung. *Dissertation.* Universität Berlin.

Bravais, A. (1847). Analyse mathématique sur les probabilités des erreurs de situation d'un point. *Mémoires présentés par divers savants à l'Académie Royale des Sciences de l'Institut de France, 9*, 255–332.

Brüel & Kjaer. (1993). *Sound Intensity.* Naerum (Denmark): Brüel & Kjaer.

BS 6841. (1987). Measurement and evaluation of human exposure to whole-body mechanical vibration and repeated shock. *Britisch Standars Institution.*

Cabrera, J., Ortiz, A., Carabias, E., & Simon, A. (2004). An Alternative Method to Determine the Magic Tyre Model Parameters Using Genetic Algorithms. *Vehicle System Dynamics, 41*(2), 109-127.

Chen, D., & Crolla, D. (1998). Subjective and objective Measures of vehicle handling: Drivers & Experiments. *Vehicle System Dynamics Supplement, 28*, 576-597.

Christ, E. (1973). Beanspruchung und Leistungsfähigkeit des Menschen bei unterbrochener und Langzeit-Exposition mit stochastischen Schwingungen. *VDI Fortschritt-Berichte, Reihe 11*(17).

Clark, S. (1977). Geometric Effects on the Rolling Resistance of Pneumatic Tires. In E. R. Committee, *Tire Rolling Losses and Fuel Economy - an R&D Planning Workshop* (pp. 111-119). Cambridge (MA): Society of Automotive Engineers.

Clark, S. (1978). Rolling Resistance of Pneumatic Tires. *Tire Science and Technology, 6*(3), 163-175.

Clark, S. (1981). *The Mechanics of Pneumatic Tires.* University of Michigan: U.S. Department of Transportation, National Highway Traffic Safety Administration.

Cohen, J. (1988). *Statistical Power Analysis for the Behavioral Sciences.* Hillsdale: Associates, Lawrence Erlbaum.

Continental. (2003). *Tyre Model Performance Test.*

Cooley, J., & Tukey, J. (1965). An Algorithm for the Machine Computation of the Complex Fourier Series. *Mathematics of Computation, 19,* 297-301.

Coulomb, C. (1785). Théories des Machines Simples. *Mémoire de Mathématique et de Physique de l'Académie Royal.*

Cucuz, S. (1993). Schwingempfindung von Pkw-Insassen - Auswirkung von stochastischen Unebenheiten und Einzelhindernissen der realen Fahrbahn. *Dissertation.* Technische Universität Braunschweig.

Davis, L. (1991). *The Handbook of Genetic Algorithms.* New York: Van Nostrand Reinhold.

Decker, M. (2009). Zur Beurteilung der Querdynamik von Personenkraftwagen. *Dissertation.* Technische Universität München.

Denker, D. (1988). Niedrige Fahrzeug-Außengeräusche und hohe Fahrsicherheit — ein Zielkonflikt? *Fortschritte der Fahrzeugtechnik, 2,* 204-218.

Deppermann, K.-H. (1989). Fahrversuche und Berechnungen zum Geradeauslauf von Personenkraftwagen. *Fortschritt-Berichte VDI, Reihe 12, Verkehrstechnik, Fahrzeugtechnik*(133).

DeRaad, L. (1977). The Influence of Road Surface Texture on Tire Rolling Resistance. In E. R. Committee, *Tire Rolling Losses and Fuel Economy - an R&D Planning Workshop* (pp. 143-149). Cambridge (MA): Society of Automotive Engineers.

Dettki, F. (2005). Methoden zur objektiven Bewertung des Geradeauslaufs von Personenkraftwagen. *Dissertation*. Universität Stuttgart.

Di Luise, P. (2015). Modelling and Analysis of the Influence Tyres have on Ride Comfort. *Supervised master thesis*. University of Naples.

Diana, G., & Cheli, F. (2005). *Cinematica e dinamica dei sistemi multicorpo.* Milano: Schönenfeld & Ziegler.

Dibbern, K. (1992). Ermittlung eines Kennwertes für den ISO-Fahrspurwechsel in Versuch und Simulation. *Fortschritt-Berichte VDI, Reihe 12, Verkehrstechnik, Fahrzeugtechnik*(164).

DIN 45680. (1997). Messung und Bewertung tieffrequenter Geräuschimmissionen in der Nachbarschaft.

DIN EN 61672-1. (2003). Elektroakustik — Schallpegelmesser.

Dobrzynski, W. (1983). Zur Bedeutung von Strömungswechseldrücken auf der Karosserieoberfläche für den Innenlärm von Personenkraftfahrzeugen. *Dissertation*. Technische Universität Berlin.

Dugoff, H., Fancher, P., & Segel, L. (1970). An analysis of tire traction properties and their influence on vehicle dynamics performance. *SAE Technical Paper 700377.*

Dupuis, H., & Hartung, E. (1972). Reqelloser und Sinusförmniger Schwingungen. *Ergonomics, 15*(3), 237-265.

Dupuis, H., & Zerlett, G. (1984). Beanspruchung des Menschen durch mechanische Schwingungen. Forschungsbericht Ganz-Körper-Schwingungen: Kenntnisstand zur Wirkung von Ganz-Körper-Schwingungen. *Hauptverband der Gewerblichen Berufsgenossenschaften.*

ERDA Rolling Resistance Advisory Committee. (1977). *Tire Rolling Losses and Fuel Economy - an R&D Planning Workshop.* Cambridge (MA): Society of Automotive Engineers.

European Union. (2009). On the labelling of tyres with respect to fuel efficiency and other essential parameters. *Regulation (EC) No 1222/2009*, L342/46-L342/58.

Evans, I. (1960). The Rolling Resistance of a Wheel with a Solid Rubber Tire. *Rubber Chemistry and Technology, 33*(2), 302-305.

Farrer, D. (1993). An Objective Measurement Technique for the Quantification of On-Centre Handling Quality. *SAE Technical Paper 930827.*

Faure, H. (1982). Discrepance de suites associees a un systeme de numeration (en dimension s). *Acta Arithmetica, 41*(4), 337-351.

Fletcher, H., & Munson, W. (1933). Loudness, its definition, measurement and calculation. *Journal of the Acoustical Society of America, 12*(4), 377-430.

Frey, H., & Patil, S. (2002). Identification and Review of Sensitivity Analysis Methods. *Risk Analysis, 22*(3), 553-578.

Frigo, M., & Johnson, S. (1998). FFTW: An Adaptive Software Architecture for the FFT. *Proceedings of the International Conference on Acoustics, Speech, and Signal Processing, 3*, S. 1381-1384.

Fuchs, J. (1993). Beitrag zum Verhalten von Fahrer und Fahrzeug bei Kurvenfahrt. *Fortschritt-Berichte VDI, Reihe 12, Verkehrstechnik, Fahrzeugtechnik*(184), 84-89.

Fujikawa, T., Funazaki, A., & Yamazaki, S. (1994). Tire tread temperature in actual contact areas. *Tire Science and Technology, 22*(1), 19-41.

Fülbier, K. (2001). Systemansatz zur Untersuchung und Beurteilung des Abrollkomforts von Kraftfahrzeugen bei der Überfahrt von Einzelhindernissen. *Dissertation*. RWTH Aachen.

Gallrein, A., & Bäcker, M. (2007). CDTire: a tire model for comfort and durability applications. *Vehicle System Dynamics, 45*, 69-77.

Galvao, L., Pizarro, M., & Epiphanio, J. (2001). Variations in Reflectance of Tropical Soils: Spectral-Chemical Composition Relationships from AVIRIS data. *Remote Sensing of Environment, 75*(2).

Gauterin, F. (2010). Reifen-Fahrbahngeräusche. In K. Genuit, *Sound-Engineering im Automobilbereich* (S. 270-279). Heidelberg: Springer-Verlag Berlin Heidelberg.

Gauterin, F., & Ropers, C. (2005). Modal tyre models for road noise improvement. *Vehicle System Dynamics, 43*, 297-304.

Genuit, K. (2010). *Sound-Engineering im Automobilbereich.* Heidelberg: Springer-Verlag Berlin Heidelberg.

Gies, S., & Marusic, Z. (2000). Das Lenkgefühl - Merkmale der subjektiven und objektiven Beschreibung. In K. Becker, *Subjektive Fahreindrücke sichtbar machen.* Renningen-Malmsheim: Expert Verlag.

Gipser, M. (1987). DNS-Tire, ein dynamisches, räumliches nichtlineares Reifenmodell. *Tagung Reifen, Fahrwerk, Fahrbahn* (S. 115–135). Düsseldorf: VDI Verlag.

Gipser, M. (1998). Reifenmodelle für Komfort- und Schlechtwegsimulationen. *Aachener Kolloquium Fahrzeug- und Motorentechnik.* RWTH Aachen: VDI Verlag.

Gipser, M. (2007). FTire - the tire simulation model for all applications related to vehicle dynamics. *Vehicle System Dynamics, 45*, 139-151.

Gobbi, M., Mastinu, G., & Miano, C. (2005). *Optimal Design of Complex Mechanical Systems.* Milan: Springer.

Greiner, M., & Heimann, P. (2013). Reifenrollwiderstand - Nassgriff - Reifen-Fahrbahn-Geräusch. *Abschlussbericht.* Universität Stuttgart, Karlsruher Institut für Technologie.

Griffin, M. (2012). *Handbook of Human Vibration.* London: Academic Press Limited.

Griffin, M., & Fothergill, L. (1977). The Evaluation of discomfort produced by multiple frequency whole-body vibration. *Ergonomics, 20*(3), 263-276.

Grollius, S. (2013). Analyse des gekoppelten Systems Reifen-Hohlraum-Rad-Radführung im Rollzustand und Entwicklung eines Rollgeräuschmodells. *Dissertation.* Karlsruher Institut für Technologie: KIT Scientific Publishing.

Guiggiani, M. (2014). *The Science of Vehicle Dynamics.* Heidelberg: Springer.

Gutjahr, D. (2014). Objektive Bewertung querdynamischer Reifeneigenschaften im Gesamtfahrzeugversuch. *Dissertation.* Karlsruher Institut für Technologie: KIT Scientific Publishing.

Hadekel, R. (1952). The mechanical characteristics of pneumatic tyres; a digest of present knowledge. *S&T Memo*(10).

Halton, J. (1960). On the efficiency of certain quasirandom sequences of points in evaluating multidimensional integrals. *Numerische Mathematik, 2*(1), 84-90.

Han, S.-O. (2011). Varianzbasierte Sensitivitätsanalyse als Beitrag zur Bewertung der Zuverlässigkeit adaptronischer Struktursysteme. *Dissertation*. Universität Darmstadt.

Harrer, M. (2007). Characterisation of Steering Feel. *Dissertation*. University of Bath.

Harris, C., & Crede, C. (1976). *Shock and Vibration Handbook.* New York: McGraw-Hill.

Hauke, J., & Kossowski, T. (2011). Comparison of Values of Pearson's and Spearman's Correlation Coefficients on the Same Sets of Data. *Quaestiones Geographicae, 30*(2), 87-93.

Heinrichmöller, D., Benner, M., & Eckstein, L. (2014). Semi-physikalisches Reifenmodell zur echtzeitfähigen Analyse. *ATZ, 116*(6), 20-25.

Heißing, B., & Brandl, H. (2002). *Subjektive Beurteilung des Fahrverhaltens.* Würzburg: Vogel Buchverlag.

Heißing, B., & Ersoy, M. (2007). *Fahrwerkhandbuch.* Wiesbaden: Vieweg & Sohn Verlag.

Heißing, B., & Ersoy, M. (2011). *Chassis Handbook: Fundamentals, Driving Dynamics, Components, Mechatronics, Perspectives.* Germany: Vieweg + Teubner Verlag.

Helfer, M. (2010). Bedeutung der Umströmungsgeräusche für das Innen- und Außengeräusch von Kraftfahrzeugen. In K. Genuit, *Sound-Engineering im Automobilbereich* (S. 279-305). Heidelberg: Springer-Verlag Berlin Heidelberg.

Hennecke, D. (1995). Zur Bewertung des Schwingungskomforts von Pkw bei instationären Anregungen. *VDI Fortschritt-Berichte, Reihe 12*(237).

Henze, R. (2004). Beurteilung von Fahrzeugen mit Hilfe eines Fahrermodells. *Dissertation*. Institut ür Fahrzeugtechnik Braunschweig.

Hieronimus, K. (1990). Anforderungen an Schwingungs- und Akustikberechnungen aus Sicht der Fahrzeugentwicklung. *VDI-Berichte*(186), 705-734.

Hilscher, C. (2008). Komfortrelevante Charakterisierung des Übertragungsverhaltens von Reifen in Messung und Simulation. *Dissertation*. Technische Universität Dresden.

Hirschberg, W., Rill, G., & Weinfurter, H. (2007). Tire model TMeasy. *Vehicle System Dynamics, 45*, 101–119.

Holt, W., & Wormeley, P. (1922). Power Losses in Automobile Tires. *Technologic Papers of the Bureau of Standards*, 451-461.

Huneke, M. (2012). Fahrverhaltensbewertung mit anwendungsspezifischen Fahrdynamikmodellen. *Dissertation*. Technische Universität Braunschweig.

ISO 226. (2003). Acoustics — Normal equal-loudness-level contours. *International Organization for Standardisation*.

ISO 2631. (1997). Mechanical vibration and shock — Evaluation of human exposure to whole-body vibration. *International Organization for Standardisation*.

ISO 28580. (2009). Passenger car, truck and bus tyres — Methods of measuring rolling resistance — Single point test and correlation of measurement results. *International Organization for Standardisation.*

ISO 362. (2009). Acoustics — Measurement of noise emitted by accelerating road vehicles — Engineering method. *International Organization for Standardisation.*

ISO 3888-2. (2011). Passenger cars — Test track for a severe lane-change manoeuvre — Part 2: Obstacle avoidance. *International Organization for Standardisation.*

Jianmin, G., Gall, R., & Zuomin, W. (2001). Dynamic Damping and Stiffness Characteristics of the Rolling Tire. *Tire Science and Technology, 29*(4), 258-268.

Johnson, M., Moore, L., & Ylvisaker, D. (1990). Minimax and Maximin Distance Designs. *Journal of Statistical Inference and Planning, 26*, 131-148.

Joseph, V., & Hung, Y. (2008). Orthogonal-Maximin Latin Hypercube Designs. *Statistica Sinica, 18*, 171-186.

Kindt, P., Gonzalez Diaz, C., Vercammen, S., Middelberg, J., Kimble, B., & Leyssens, J. (2013). Effects of Rotation on the Tire Dynamic Behaviour: Experimental and Numerical Analyses. *Tire Science and Technology, 41*(4), 248-261.

Kleppmann, W. (2008). *Taschenbuch Versuchsplanung: Produkte und Prozesse optimieren.* München: Carl Hanser Verlag.

Kudritzki, D. (1989). Zum Einfluß querdynamischer Bewegungsgrößen auf die Beurteilung des Fahrverhaltens. *Fortschritt-Berichte VDI, Reihe 12, Verkehrstechnik, Fahrzeugtechnik*(132).

LaClair, T. (2006). Rolling Resistance. In *The Pneumatic Tire* (pp. 476-532). U.S. Department of Transportation, National Highway Traffic Safety Administration (NHTSA).

Landwehr, R. (2013). Reifen Label, Ökologie und Sicherheit. *ATZ, 115*(7-8), 566-571.

Mastinu, G., Gobbi, M., & Miano, C. (2004). *Multi-Objective Optimization of the Handling Performances of a Road Vehicle: A Fundamental Study on Tire Selection.* Politecnico di Milano, Department of Mechanical Engineering. Milan: ASME, Journal of Mechanical Design.

Matousek, J. (1998). On the L2-Discrepancy for Anchored Boxes. *Journal of Complexity, 14*(4), 527-556.

Maurice, J. (2000). Short wavelength and dynamic tyre behaviour under lateral and combined slip conditions. *Dissertation.* Delft University of Technology.

McKay, M., Beckman, R., & Conover, W. (1979). A comparison of three methods for selecting values of input variables in the analysis of output from a computer code. *Technometrics, 21*(2), 239-245.

Michelin. (2005). *The Tyre.* Clermont-Ferrant: Société de Technologie Michelin.

Mitschke, M., & Wallentowitz, H. (2003). *Dynamik der Kraftfahrzeuge.* Braunschweig: Springer-Verlag Berlin Heidelberg New York.

Mitschke, M., Cucuz, S., & Hennecke, D. (1995). Bewertung und Summationsmechanismen von ungleichmässig regellosen Schwingungen. *ATZ, 97*(11), 784-791.

Morokoff, W., & Caflisch, R. (1994). Quasi-Random Sequences and their Discrepancies. *Society for Industrial and Applied Mathematics, 16*(5), 1251-1279.

MSC.Software Corporation. (2003). *Using ADAMS/Tire Tire Models.* USA: MSC.Software Corporation.

Niedermeier, F. (2015). Virtuelle Grundauslegung funktionaler Reifeneigenschaften im Auslegungsprozess Fahrdynamik. *Dissertation*. Karlsruher Institut für Technologie: KIT Scientific Publishing.

Niedermeier, F., Peckelsen, U., & Gauterin, F. (2013). Virtual Tires in the Early Vehicle Development Stage. *International Journal of Emerging Technology and Advanced Engineering, 3*(7), 394-401.

Niederreiter, H. (1988). Low Discrepancy and Low Dispersion Sequences. *Journal of Number Theory, 31*(1), 51-70.

Niederreiter, H. (1992). *Random Number Generation and Quasi-Monte Carlo Methods.* Philadelphia: Society for Industrial and Applied Mathematics.

Niemeyer, P., Eckstein, L., Kessen, H., Klein, M., & Wegener, D. (2011). Experimentelle Ermittlung hochdynamischer Reifeneigenschaften. Messungen am stehenden und rollenden Reifen bei vertikalen Anregungen bis 25 Hz. *VDI-Berichte*(2137), 79-98.

Oertel, C., & Fandre, A. (2001). Das Reifenmodellsystem RMOD-K. *ATZ*(11), 1074-1079.

Olley, M. (1947). Road Manners of the Modern Car. *Proceedings of the Institution of Automobile Engineers*, 147 - 161.

Pacejka, H. (1958). *Study of the lateral behaviour of an automobile moving upon a flat level road.* Cornell Aeronautical Laboratory Report YC-857-F-23.

Pacejka, H. (2006). *Tyre and Vehicle Dynamics.* Oxford: Butterworth-Heinemann.

Pacejka, H., & Bakker, E. (1992). The Magic Formula tyre model. *Vehicle System Dynamics, 21*, 1-18.

Parson, A., & Kraemer, S. (1987). Zum Problem der subjektiven Bewertung des Reifenkomforts. *VDI-Berichte*(650).

Pearson, K. (1896). Mathematical contributions to the theory of evolution. III. Regression, heredity, and panmixia. *Philosophical Transactions of the Royal Society of London, Series A, 187*, 253–318.

Peckelsen, U. (2012). Analysis and Optimisation of Tires for Passenger Cars. *Master thesis*. Politecnico di Milano.

Peckelsen, U., & Gauterin, F. (2013). Influence of Driving Conditions on Total Rolling Resistance. *ATZ, 115*(11), 86-91.

Peckelsen, U., D'Avanzo, T., Bode, D., Brenker, M., & Gauterin, F. (2013). Rolling Resistance: a Comprehensive Study under Real Vehicle-Driving Conditions. *3rd International Munich Chassis Symposium* (S. 709-728). Munich: Springer Verlag.

Peckelsen, U., Gauterin, F., & Unrau, H.-J. (2015). Definition and Modeling of the Power Loss of Tires. *Tire Science and Technology, 43*(2), 163-180.

Pletschen, B. (2010). Akustik in der Fahrzeugentwicklung. In K. Genuit, *Sound-Engineering im Automobilbereich* (S. 89-108). Heidelberg: Springer-Verlag Berlin Heidelberg.

Pottinger, M., & Yager, T. (1986). *The Tire Pavement Interface: A Symposium.* Baltimore: Astm International.

Radt, H., & Milliken, W. (1983). Non-dimensionalizing Tyre Data for Vehicle Simulation. *Road Vehicle Handling Inst. of Mechanical Engineers Publications*, 229-240.

Redlich, P. (1994). *Objektive und subjektive Beurteilung aktiver Vierradlenkstrategien.* Aachen: Shaker Verlag.

Reichelt, W., & Strackerjan, B. (1992). Bewertung der Fahrdynamik vom Pkw im geschlossenen Regelkreis mit Hilfe von Fahrsimulatoren und Fahrermodellen. *VDI Berichte*(948), 251-273.

Rericha, I. (1986). Methoden zur objektiven Bewertung des Fahrkomforts. *Automobil-Industrie*, 175-182.

Reynolds, O. (1874). On Rolling-Friction. *The Engineer*, 155-174.

Rhyne, T., & Cron, S. (2012). A Study on Minimum Rolling Resistance. *Tire Science and Technology, 40*(4), 220–233.

Riedel, A., & Arbinger, R. (1997). Subjektive und objektive Beurteilung des Fahrverhaltens von Pkw. *FAT-Schriftenreihe, 139*.

Riegel, M., & Wiedemann, J. (2003). Bestimmung des Windgeräuschanteils im Vergleich zu Antriebs- und Rollgeräusch im Innenraum von Pkw. In M. Bargende, & J. Wiedemann, *5. Internationales StuttgarterSymposium Kraftfahrwesen und Verbrennungsmotoren 18.-20.2.2003.* Stuttgart: Renningen: Expert-Verlag.

Riegel, M., & Wiedemann, J. (2008). Messung des Reifen-Fahrbahn-Geräusches im Innenraum eines Pkw. *ATZ, 110*(9), 822-828.

Robinson, D., & Dadson, R. (1956). A re-determination of equal-loudness relations for pure tones. *Britisch Journal of Applied Physics, 7*(5), 166–181.

Rompe, K., & Ehlich, L. (1978). Zum Stand der objektiven Bewertungen von Reifen- und Fahrwerkseigenschaften. *VDI Berichte 650.*

Rönitz, R., Braess, H.-H., & Zomotor, A. (1977). Verfahren und Kriterien zur Bewertung des Fahrverhaltens von Personenkraftwagen - Stand und Problematik - Teil I+II. *Automobil Industrie, 77*(1), 29-39.

Sammet, T. (2007). Motion-Cueing-Algorithmen für die Fahrsimulation. *Dissertation*. Technische Universität München.

Sandberg, U., & Ejsmont, J. (2002). *Tyre/Road Noise Reference Book.* Harg (Sweden): INFORMEX Ejsmont & Sandberg Handelsbolag.

Schimmel, C. (2010). Entwicklung eines fahrerbasierten Werkzeugs zur Objektivierung subjektiver Fahreindrücke. *Dissertation*. Technische Universität München.

Schmeitz, A. (2004). A semi-empirical three-dimensional model of the pneumatic tyre rolling over arbitrary uneven road surfaces. *Dissertation*. Delft University of Technology.

Schmeitz, A., Besselink, I., & Jansen, S. (2007). TNO MF-SWIFT. *Vehicle System Dynamics, 45*(1), 121-137.

Schulze, T., Bolz, G., Strübel, C., & Wies, B. (2010). Reifen im Zielkonflikt zwischen Rollwiderstand und Nassgriff. *ATZ, 112*(7/8), 516-523.

Schuring, D. (1976). Energy Loss of Pneumatic Tires Under Freely Rolling, Braking, and Driving Conditions. *Tire Science and Technology, 4*(1), 3-15.

Schuring, D. (1977). A New Look at the Definition of Tire Rolling Loss. In E. R. Committee, *Tire Rolling Losses and Fuel Economy - an R&D Planning Workshop* (pp. 31-37). Cambridge (MA): Society of Automotive Engineers.

Siebertz, K., van Bebber, D., & Hochkirchen, T. (2010). *Statistische Versuchsplanung: Design of Experiments (DoE).* Heidelberg: Springer-Verlag.

Simic, D. (1970). Beitrag zur Optimierung der Schwingungseigenschaften des Fahrzeuges - Physiologische Grundlagen des Schwingungskomforts. *Dissertation*. Technische Universität Berlin.

Sobol, I. (1967). On the distribution of points in a cube and the approximate evaluation of integrals. *U.S.S.R. Computational Mathematics and Mathematical Physics, 7*(4), 86-112.

Sobol, I. (1993). *Sensitivity analysis for nonlinear mathematical models.* Mathematical Modelling and Computational Experiment.

Sochor, D. (2014). Qualitative Bewertung von dynamischen Radnabenkräften bei Fahrbahnanregung. *Diplomarbeit*. Hochschule für angewandte Wissenschaften München.

Spearman, C. (1904). The Proof and Measurement of Association between Two Things. *The American Journal of Psychology, 15*(1), 72-101.

Stiber, N., Pantazidou, M., & Small, M. (1999). Expert System Methodology for Evaluating Reductive Dechlorination at TCE Sites. *Environmental Science & Technologies, 33*(17), 3012-3020.

Strange, R. (1982). Instrumented Objective Tire/Vehicle Handling Testing. *SAE Technical Paper 820456.*

Tischleder, J., Köhne, S., Leister, G., & Bode, D. (2004). History and current status of the TIME-procedure to measure force and moments properties of tires. *Tire.Wheel.Tech.* Munich.

TNO Automotive. (2001). *MF-Tyre User Manual Version 5.2.* Netherlands: TNO Automotive.

TNO Automotive. (2010). *MF-Tyre/MF-SWIFT 6.1.2 Equation Manual.* Netherlands: TNO Automotive.

Tomaske, W. (1983). Einfluß der Bewegungsinformation auf das Lenkregelverhalten des Fahrers sowie Folgerungen für die Auslegung von Fahrsimulatoren. *Dissertation.* Hochschule der Bundeswehr Hamburg.

Trosset, M. (1999). Approximate Maximin Distance Designs. In *Proceedings of the Section on Physical and Engineering Sciences* (S. 223-227). Williamsburg: American Statistical Association.

Troulis, M. (2002). Übertragungsverhalten von Radaufhängungen für Personenwagen im komfortrelevanten Frequenzbereich. *Dissertation.* Technische Hochschule Karlsruhe.

VDI 2057 Blatt 1. (2002). Einwirkung mechanischer Schwingungen auf den Menschen - Ganzkörper-Schwingungen. *VDI-Richtlinie.*

VDI 2057 Blatt 1. (2002). Einwirkung mechanischer Schwingungen auf den Menschen - Ganzkörper-Schwingungen. *VDI-Richtlinie.*

Venneborger, M., Strübel, C., Wies, B., & Wiese, K. (2013). Leichtlaufreifen für PKW mit niedrigem CO2-Ausstoss. *ATZ, 115*(7-8), 572-577.

Weir, H., & DiMarco, R. (1978). Correlation and Evaluation of Driver/Vehicle Directional Handling Data. *SAE Technical Paper 780010.*

Welch, P. (1967). The Use of Fast Fourier Transform for the Estimation of Power Spectra: A Method Based on Time Averaging Over Short, Modified Periodograms. *IEEE Transactions on Audio Electroacoustics, AU-15*, 70-73.

Wolf, H. (2009). Ergonomische Untersuchung des Lenkgefühls an Personenkraftwagen. *Dissertation*. Technische Universität München.

Wormeley, P., & Holt, W. (1922). Power Losses in Automobile Tires. *Technologic Papers of the Bureau of Standards*, 451-461.

Youngblut, C., Johnston, R., Nash, S., Wienclaw, R., & Will, C. (1996). Review of Virtual Environment. *IDA Paper P-3186*. Institute for Defense Analyses (Virginia).

Zegelaar, P. (1998). The dynamic response of tyres to brake torque variations and road unevennesses. *Dissertation*. Delft University of Technology.

Zeller, P. (2012). *Handbuch Fahrzeugakustik.* Wiesbaden: Vieweg+Teubner Verlag.

Zomotor, A., Braess, H., & Rönitz, R. (1997). Verfahren und Kriterien zur Bewertung des Fahrverhaltens von Personenkraftwagen, Ein Rückblick auf die letzten 20 Jahre, Teil 1. *ATZ, 99*(12), 780-786.

Zomotor, A., Braess, H., & Rönitz, R. (1998). Verfahren und Kriterien zur Bewertung des Fahrverhaltens von Personenkraftwagen, Ein Rückblick auf die letzten 20 Jahre, Teil 2. *ATZ, 100*(3), 236-243.

Zschocke, A. (2009). Ein Beitrag zur objektiven und subjektiven Evaluierung des Lenkkomforts von Kraftfahrzeugen. *Dissertation*. Universität Karlsruhe.

List of Figures

List of Tables

List of Abbreviations

ABS	Anti-lock Brake Systems
ANN	Artificial Neural Network
ARS	Anti-Roll System
ASR	Anti-Slip Regulation
BRIT	Brush and Ring Tyre
CD-Tyre	Comfort and Durability Tyre
C-Tyre	Comfort Tyre
DFT	Discrete Fourier Transform
DNS	Dynamical Non-Linear Spatial
DoB	Degrees of Boundaries
DoE	Design of Experiments
DoF	Degrees of Freedom
ESP	Electronic Stability Programme
FEM	Finite Element Models
FFD	Full-Factorial Design
FFT	Fast Fourier Transform
FRF	Frequency Response Function
FrFD	Fractional-Factorial Design
FTC	Functional Tyre Characteristic
F-Tyre	Flexible Ring Tyre
HS	Halton Sequence
HSS	Halton Sequence with Scramble
LDF	Lateral Dynamics Formulae
LDS	Low Discrepancy Sequences
LHS	Latin Hypercube Sampling
MF	Magic Formula
NVH	Noise Vibration Harshness
OEM	Original Equipment Manufacturer
OMC	Objective Manoeuvre Criterion

RFT	Run-Flat Tyre
RMOD-K	Reifen Modell Komfort
RS	Random Sequences
SMI	Subjective Manoeuvre Index
SS	Sobol Sequence
SSS	Sobol Sequence with Scramble
STD	Standard Tyre
SWIFT	Short Wavelength Intermediate Frequency Tyre
TCS	Traction Control System
TDP	Tyre Design Parameter
TES	Tyre Enveloping Stiffness
TMeasy	Tyre Model easy
VDC	Vehicle Dynamic Control
VDF	Vehicle Dynamics Formulae

List of Symbols

Tyre geometrical properties

R	Tyre radius	$[m]$
R_r	Rim radius	$[m]$
$R_{y,b}$	Curvature of the belt	$[m]$
$R_{y,t}$	Curvature of the tread	$[m]$
w_n	Nominal width	$[m]$
w_s	Sidewall thickness	$[m]$
w_b	Belt width	$[m]$
w_t	Tread width	$[m]$
h_s	Sidewall height	$[m]$
h_b	Belt height	$[m]$
h_t	Tread height	$[m]$
h^*_{Gy}	Weighted height for the Calculation of the Total Shear Modulus	$[m]$
h^*_{Gy1}	Weighted height for the Calculation of the Shear Modulus of Sidewall and Inflation Pressure	$[m]$
h^*_{Gy2}	Weighted height for the Calculation of the Shear Modulus of Belt and Tread	$[m]$
e_r	Void ratio	$[-]$
e_C	Coulomb's horizontal distance between wheel hub and road obstacle	$[m]$
h_C	Coulomb's vertical distance between wheel hub and road obstacle	$[m]$
ξ	Normalised coordinate describing the longitudinal position along the contact path	$[-]$
ξ_a	Critical normalised coordinate describing the longitudinal position along the contact path	$[-]$

Displacements

z_h	Vertical displacement of the wheel hub	$[m]$
z_r	Vertical displacement of the rim	$[m]$
z_b	Vertical displacement of the belt	$[m]$
$\tilde{z}_h$	Vertical displacement of the wheel hub at static tyre load	$[m]$
$\tilde{z}_r$	Vertical displacement of the rim at static tyre load	$[m]$
$\tilde{z}_b$	Vertical displacement of the belt at static tyre load	$[m]$
$\bar{\gamma}$	Shear strain	$[-]$
γ_k	Kinematic shear strain	$[-]$
γ_a	Shear strain for adhesion conditions	$[-]$
γ_s	Shear strain for sliding conditions	$[-]$

Areas, volumes and moments of inertia

w_c	Width of contact area	$[m]$
l_c	Length of contact area	$[m]$
w_p	Width of the area on which inflation pressure works	$[m]$
l_p	Length of the area on which inflation Pressure works	$[m]$
$A_{s,xy}$	Sidewall area	$[m^2]$
$A_{p,xy}$	Area on which inflation pressure works	$[m^2]$
$A_{c,xy}$	Contact area	$[m^2]$
A^*_{Gy}	Weighted area for the Calculation of the Total Shear Modulus	$[m]$
A^*_{Gy1}	Weighted area for the Calculation of the Shear Modulus of Sidewall and Inflation Pressure	$[m]$
A^*_{Gy2}	Weighted area for the Calculation of the Shear Modulus of Belt and Tread	$[m]$
V_s	Volume of sidewall (subjected to hysteresis)	$[m^3]$
V_b	Volume of belt (subjected to hysteresis)	$[m^3]$

V_t	Volume of tread (subjected to hysteresis)	$[m^3]$
$I_{s,yx}$	Moment of inertia of sidewall	$[m^4]$
$I_{s,zy}$	Moment of inertia of sidewall	$[m^4]$
$I_{b,zy}$	Moment of inertia of belt	$[m^4]$
$I_{t,zy}$	Moment of inertia of tread	$[m^4]$

Tyre material properties

ε_0	Strain amplitude	$[-]$
$\varepsilon_{s,0}$	Strain amplitude of sidewall	$[-]$
$\varepsilon_{b,0}$	Strain amplitude of belt	$[-]$
$\varepsilon_{t,0}$	Strain amplitude of tread	$[-]$
σ_0	Stress amplitude	$[N/m^2]$
$\sigma_{s,0}$	Stress amplitude of sidewall	$[N/m^2]$
$\sigma_{b,0}$	Stress amplitude of belt	$[N/m^2]$
$\sigma_{t,0}$	Stress amplitude of tread	$[N/m^2]$
δ	Phase angle between stress and strain	$[rad]$
E_s	Elastic modulus of the sidewall	$[N/m^2]$
E_b	Elastic modulus of the belt	$[N/m^2]$
E_t	Elastic modulus of the tread	$[N/m^2]$
G_y	Total shear modulus	$[N/m^2]$
G_{y1}	Shear modulus of sidewall and due to inflation pressure	$[N/m^2]$
G_{y2}	Shear modulus of belt and tread	$[N/m^2]$
G_{ys}	Shear modulus of sidewall	$[N/m^2]$
G_{yp}	Shear modulus due to inflation pressure	$[N/m^2]$
G_{yb}	Shear modulus of belt	$[N/m^2]$
G_{yt}	Shear modulus of tread	$[N/m^2]$
k	Total vertical stiffness	$[N/m]$
k_{y1}	Lateral stiffness of sidewall and due to inflation pressure	$[N/m]$
k_{y2}	Lateral stiffness of belt and tread	$[N/m]$

k_{z1}	Vertical stiffness of sidewall and due to inflation pressure	$[N/m]$
k_{z2}	Vertical stiffness of belt and tread	$[N/m]$
k_r	Rim stiffness	$[N/m]$
k_a	Air stiffness	$[N/m]$
k_s	Sidewall stiffness	$[N/m]$
k_p	Vertical stiffness due to inflation pressure	$[N/m]$
k_b	Belt stiffness	$[N/m]$
k_t	Tread stiffness	$[N/m]$
d_{y1}	Lateral damping of sidewall and due to inflation pressure	$[Ns/m]$
d_{y2}	Lateral damping of belt and tread	$[Ns/m]$
d_{z1}	Vertical damping of sidewall and due to inflation pressure	$[Ns/m]$
d_{z2}	Vertical damping of belt and tread	$[Ns/m]$
d_r	Rim damping	$[Ns/m]$
d_a	Air damping	$[Ns/m]$
d_s	Sidewall damping	$[Ns/m]$
d_b	Belt damping	$[Ns/m]$
d_t	Tread damping	$[Ns/m]$
m_r	Rim mass	$[kg]$
m_a	Cavity air mass	$[kg]$
m	Tyre mass	$[kg]$
ρ_a	Density of the air	$[kg/m^3]$

Road properties

z_e	Road excitation	$[m]$
$\bar{z}_e$	Road excitation amplitude	$[m]$
f_e	Road excitation frequency	$[Hz]$
μ_a	Distribution of adhesion coefficient	$[-]$
μ_s	Distribution of sliding coefficient	$[-]$
$\mu_{a,ref}$	Reference adhesion coefficient	$[-]$
$\mu_{s,ref}$	Reference sliding coefficient	$[-]$

Form factors for physical tyre model

f_{k_s}	Form factor 1	$[-]$
f_{k_p}	Form factor 2	$[-]$
$f_{G_{ys}}$	Form factor 3	$[-]$
f_{G_y}	Form factor 4	$[-]$
f_{d_y}	Form factor 5	$[-]$
f_{d_z}	Form factor 6	$[-]$
q_1	Exponent for pressure distribution	$[-]$
q_2	Exponent for friction coefficient distribution	$[-]$
q_3	Ratio between reference adhesion and sliding coefficient	$[-]$

Tyre test bench operating conditions and moments

θ	Rotation angle	$[rad]$
Ω	Angular velocity	$[rad/s]$
M_{drum}	Drum torque	$[Nm]$

Tyre operating conditions, forces and moments

p	Inflation pressure	$[Pa]$
κ	Longitudinal slip	$[-]$
α	Slip angle	$[rad]$
γ	Camber angle	$[rad]$
ϑ	Rotation angle	$[rad]$
ω	Angular velocity	$[rad/s]$
v_x	Trajectory velocity of the tyre	$[m/s]$
F_x	Longitudinal force (parallel to the road)	$[N]$
F_y	Lateral force (parallel to the road)	$[N]$
F_z	Vertical force (also tyre load)	$[N]$
μ_x	Longitudinal friction coefficient	$[-]$
μ_y	Lateral friction coefficient	$[-]$
M_x	Overturning torque	$[Nm]$
M_y, M_r	Rolling resistance torque	$[Nm]$
M_z	Aligning torque	$[Nm]$

t	Pneumatic trail	$[m]$
$M_{s,B}$	Bending torque due to sidewall	$[Nm]$
$M_{b,B}$	Bending torque due to belt	$[Nm]$
$M_{t,B}$	Bending torque due to tread	$[Nm]$
F_{aero}	Aerodynamic force	$[N]$
M_{aero}	Aerodynamic torque	$[Nm]$

Vehicle operating conditions

φ	Roll angle	$[rad]$
ϑ	Pitch angle	$[rad]$
ψ	Yaw angle	$[rad]$
β	Sideslip angle	$[rad]$
v	Driving velocity	$[m/s]$
x_s	Longitudinal displacement of sprung masses	$[m/s^2]$
y_s	Lateral displacement of sprung masses	$[m/s^2]$
z_s	Vertical displacement of sprung masses	$[m/s^2]$
x_u	Longitudinal displacement of unsprung masses	$[m/s^2]$
y_u	Lateral displacement of unsprung masses	$[m/s^2]$
z_u	Vertical displacement of unsprung masses	$[m/s^2]$
a_x	Longitudinal acceleration	$[m/s^2]$
a_y	Lateral acceleration	$[m/s^2]$
a_z	Vertical (bouncing) acceleration	$[m/s^2]$
$a_{...,s}$	Acceleration of sprung masses	$[m/s^2]$
$a_{...,u}$	Acceleration of unsprung masses	$[m/s^2]$

Steering characteristics

δ_H	Steering wheel angle	$[rad]$
$f_{H,max}$	Steering wheel angle amplitude	$[rad]$
f_{δ_H}	Steering wheel angle frequency	$[Hz]$
M_H	Steering wheel torque	$[Nm]$

Energy and power

ΔE_c	Variation of kinetic energy	$[J]$

ΔE	Variation of the total energy	$[J]$
ΔE_m	Variation of mechanical energy	$[J]$
ΔU	Variation of internal energy	$[J]$
W	Mechanical work	$[J]$
V	Potential energy	$[J]$
D	Damping energy	$[J]$
E_{loss}	Total energy loss	$[J]$
Q	Energy loss due to heat exchange with the environment	$[J]$
Q_r	Energy loss generated by relative displacement and rotations	$[J]$
Q_h	Energy loss generated by non-elastic deformations	$[J]$
P_{loss}	Power loss	$[W]$

Rolling resistance

t	Time	$[s]$
ω	Angular frequency	$[rad/s]$
F_r	Rolling resistance force	$[N]$
$F_{r,bend}$	Bending resistance force	$[N]$
$F_{r,air}$	Air resistance force	$[N]$
$F_{r,fr}$	Frictional resistance force	$[N]$
$F_{r,road}$	Road resistance force	$[N]$
$F_{r,slip}$	Resistance force due to slip angle	$[N]$
$F_{r,fb}$	Resistance force due to bearing friction and residual braking	$[N]$
$M_{r,0}$	Rolling resistance torque	$[Nm]$
$M'_{r,0,v}$	Variation of rolling resistance torque due to velocity	$[\%s/m]$
$M'_{r,0,F_z}$	Variation of rolling resistance torque due to tyre load	$[\%/N]$
$M'_{r,0,\kappa}$	Variation of rolling resistance torque due to longitudinal slip	$[\%/-]$

Lateral dynamics

$F_{y,0}$	Lateral force at zero slip angle	$[N]$
$K_{y,0}$	Cornering stiffness	$[N/°]$
$F_{y,max}$	Maximal lateral force	$[N]$
$\mu_{y,max}$	Maximal lateral friction	$[N/N]$
$\alpha_{F_{y,max}}$	Slip angle of the maximal lateral force	$[°]$
$\alpha_{max,\%}$	Position of maximal lateral friction	$[\%]$
$F_{y,lim}$	Lateral force at 15° slip angle	$[N]$
$\mu'_{y,max,\alpha}$	Degression of lateral force over slip angle	$[\%/°]$
$\sigma_{y,0}$	Lateral relaxation length	$[m]$
$\sigma'_{y,0,F_z}$	Variation of lateral relaxation length due to tyre load	$[\%/N]$

Ride comfort

$K_{z,0}$	Dynamic vertical stiffness at zero hertz	$[N/m]$
$K'_{z,14,f}$	Slope of dynamic vertical stiffness	$[Ns/m]$

Acoustics

L_p	Sound pressure level	$[dB]$
L_{pA}	A-weighted sound pressure level	$[dB]$
f_1, f_2, f_3, f_4	Standardised coefficients for the a-weighted sound pressure level	$[-]$
A_{1000}	Standardised coefficient for the a-weighted sound pressure level	$[dB]$
p_A	Sound pressure level	$[dB]$
$\bar{S}_L$	Objective manoeuvre criteria for low frequency noise	$[dBA]$
$\bar{S}_M$	Objective manoeuvre criteria for middle frequency noise	$[dBA]$
$\bar{S}_C$	Objective manoeuvre criteria for air cavity noise	$[dBA]$
$\bar{S}_{col}$	Objective manoeuvre criteria for acoustic colour	$[dBA]$

$\bar{F}_{x,L}$	Average longitudinal force amplitudes at low frequencies	$[N]$
$\bar{F}_{y,M}$	Average lateral force amplitudes at middle frequencies	$[N]$
$\bar{F}_{z,V}$	Average vertical force amplitudes due to first eigenmode	$[N]$
$\bar{F}_{z,M}$	Average vertical force amplitudes at middle frequencies	$[N]$
$\bar{F}_{z,C}$	Average force amplitudes due to air cavity	$[N]$
$\hat{F}_{z,V}$	Relative force peak of first vertical eigenmode	$[N]$
$\hat{F}_{z,C}$	Relative force peak due to air cavity	$[N]$

Design of experiments

n	Number of experiments (or samples)	$[-]$
n_{DV}	Number of design variables	$[-]$
d	Number of dimensions	$[-]$
D	Discrepancy	$[-]$
d_{ij}	Euclidian distance	$[-]$
H	Global homogeneity index	$[-]$
H^{doe}	Homogeneity of the DoE	$[-]$
H^{*}	Mean homogeneity along one single dimensions	$[-]$
H^{corr}	Mean correlation index	$[-]$

Correlation and sensitivity analyses

ρ	Correlation index	$[-]$
R^2	R-squared	$[-]$
S^1	Sobol's first order coefficient	$[-]$
S^{tot}	Sobol's total coefficient	$[-]$
σ	Standard deviation	$[-]$
σ^2	Variance	$[-]$
cov	Covariance	$[-]$

Karlsruher Schriftenreihe Fahrzeugsystemtechnik (ISSN 1869-6058)

Herausgeber: FAST Institut für Fahrzeugsystemtechnik

Band 1 Urs Wiesel
Hybrides Lenksystem zur Kraftstoffeinsparung im schweren Nutzfahrzeug. 2010
ISBN 978-3-86644-456-0

Band 2 Andreas Huber
Ermittlung von prozessabhängigen Lastkollektiven eines hydrostatischen Fahrantriebsstrangs am Beispiel eines Teleskopladers. 2010
ISBN 978-3-86644-564-2

Band 3 Maurice Bliesener
Optimierung der Betriebsführung mobiler Arbeitsmaschinen. Ansatz für ein Gesamtmaschinenmanagement. 2010
ISBN 978-3-86644-536-9

Band 4 Manuel Boog
Steigerung der Verfügbarkeit mobiler Arbeitsmaschinen durch Betriebslasterfassung und Fehleridentifikation an hydrostatischen Verdrängereinheiten. 2011
ISBN 978-3-86644-600-7

Band 5 Christian Kraft
Gezielte Variation und Analyse des Fahrverhaltens von Kraftfahrzeugen mittels elektrischer Linearaktuatoren im Fahrwerksbereich. 2011
ISBN 978-3-86644-607-6

Band 6 Lars Völker
Untersuchung des Kommunikationsintervalls bei der gekoppelten Simulation. 2011
ISBN 978-3-86644-611-3

Band 7 3. Fachtagung
Hybridantriebe für mobile Arbeitsmaschinen. 17. Februar 2011, Karlsruhe. 2011
ISBN 978-3-86644-599-4

Die Bände sind unter www.ksp.kit.edu als PDF frei verfügbar oder als Druckausgabe bestellbar.

Karlsruher Schriftenreihe Fahrzeugsystemtechnik (ISSN 1869-6058)

Herausgeber: FAST Institut für Fahrzeugsystemtechnik

Band 8	Vladimir Iliev **Systemansatz zur anregungsunabhängigen Charakterisierung des Schwingungskomforts eines Fahrzeugs.** 2011 ISBN 978-3-86644-681-6
Band 9	Lars Lewandowitz **Markenspezifische Auswahl, Parametrierung und Gestaltung der Produktgruppe Fahrerassistenzsysteme. Ein methodisches Rahmenwerk.** 2011 ISBN 978-3-86644-701-1
Band 10	Phillip Thiebes **Hybridantriebe für mobile Arbeitsmaschinen. Grundlegende Erkenntnisse und Zusammenhänge, Vorstellung einer Methodik zur Unterstützung des Entwicklungsprozesses und deren Validierung am Beispiel einer Forstmaschine.** 2012 ISBN 978-3-86644-808-7
Band 11	Martin Gießler **Mechanismen der Kraftübertragung des Reifens auf Schnee und Eis.** 2012 ISBN 978-3-86644-806-3
Band 12	Daniel Pies **Reifenungleichförmigkeitserregter Schwingungskomfort – Quantifizierung und Bewertung komfortrelevanter Fahrzeugschwingungen.** 2012 ISBN 978-3-86644-825-4
Band 13	Daniel Weber **Untersuchung des Potenzials einer Brems-Ausweich-Assistenz.** 2012 ISBN 978-3-86644-864-3
Band 14	**7. Kolloquium Mobilhydraulik. 27./28. September 2012 in Karlsruhe.** 2012 ISBN 978-3-86644-881-0
Band 15	4. Fachtagung **Hybridantriebe für mobile Arbeitsmaschinen 20. Februar 2013, Karlsruhe.** 2013 ISBN 978-3-86644-970-1

Die Bände sind unter www.ksp.kit.edu als PDF frei verfügbar oder als Druckausgabe bestellbar.

Karlsruher Schriftenreihe Fahrzeugsystemtechnik (ISSN 1869-6058)

Herausgeber: FAST Institut für Fahrzeugsystemtechnik

Band 16	Hans-Joachim Unrau **Der Einfluss der Fahrbahnoberflächenkrümmung auf den Rollwiderstand, die Cornering Stiffness und die Aligning Stiffness von Pkw-Reifen.** 2013 ISBN 978-3-86644-983-1
Band 17	Xi Zhang **Untersuchung und Entwicklung verschiedener Spurführungsansätze für Offroad-Fahrzeuge mit Deichselverbindung.** 2013 ISBN 978-3-7315-0005-6
Band 18	Stefanie Grollius **Analyse des gekoppelten Systems Reifen-Hohlraum-Rad-Radführung im Rollzustand und Entwicklung eines Rollgeräuschmodells.** 2013 ISBN 978-3-7315-0029-2
Band 19	Tobias Radke **Energieoptimale Längsführung von Kraftfahrzeugen durch Einsatz vorausschauender Fahrstrategien.** 2013 ISBN 978-3-7315-0069-8
Band 20	David Gutjahr **Objektive Bewertung querdynamischer Reifeneigenschaften im Gesamtfahrzeugversuch.** 2014 ISBN 978-3-7315-0153-4
Band 21	Neli Ovcharova **Methodik zur Nutzenanalyse und Optimierung sicherheitsrelevanter Fahrerassistenzsysteme.** 2014 ISBN 978-3-7315-0176-3
Band 22	Marcus Geimer, Christian Pohlandt **Grundlagen mobiler Arbeitsmaschinen.** 2014 ISBN 978-3-7315-0188-6
Band 23	Timo Kautzmann **Die mobile Arbeitsmaschine als komplexes System.** 2014 ISBN 978-3-7315-0187-9

Die Bände sind unter www.ksp.kit.edu als PDF frei verfügbar oder als Druckausgabe bestellbar.

Karlsruher Schriftenreihe Fahrzeugsystemtechnik (ISSN 1869-6058)

Herausgeber: FAST Institut für Fahrzeugsystemtechnik

Band 24 Roman Weidemann
Analyse der mechanischen Randbedingungen zur Adaption der oszillierenden Hinterschneidtechnik an einen Mobilbagger. 2014
ISBN 978-3-7315-0193-0

Band 25 Yunfan Wei
Spurführungsregelung eines aktiv gelenkten Radpaars für Straßenbahnen. 2014
ISBN 978-3-7315-0232-6

Band 26 David Schmitz
Entwurf eines fehlertoleranten Lenkventils für Steer-by-Wire Anwendungen bei Traktoren. 2014
ISBN 978-3-7315-0264-7

Band 27 Christian Schwab
Beitrag zu einer universellen Baggerschnittstelle zur Übertragung elektrischer und hydraulischer Leistung sowie elektronischer Signale für komplexe Anbaugeräte. 2014
ISBN 978-3-7315-0281-4

Band 28 Peter Dengler
Untersuchung zum effizienten Betrieb von Hydraulikzylindern in Konstantdrucksystemen unter Verwendung einer Zwischendruckleitung. 2015
ISBN 978-3-7315-0295-1

Band 29 Manuel Bös
Untersuchung und Optimierung der Fahrkomfort- und Fahrdynamikeigenschaften von Radladern unter Berücksichtigung der prozessspezifischen Randbedingungen. 2015
ISBN 978-3-7315-0310-1

Band 30 5. Fachtagung
Hybride und energieeffiziente Antriebe für mobile Arbeitsmaschinen
25. Februar 2015, Karlsruhe. 2015
ISBN 978-3-7315-0323-1

Die Bände sind unter www.ksp.kit.edu als PDF frei verfügbar oder als Druckausgabe bestellbar.

Karlsruher Schriftenreihe Fahrzeugsystemtechnik (ISSN 1869-6058)

Herausgeber: FAST Institut für Fahrzeugsystemtechnik

Band 31 Michael Eckert
Energieoptimale Fahrdynamikregelung mehrmotoriger Elektrofahrzeuge. 2015
ISBN 978-3-7315-0332-3

Band 32 Martin Scherer
Beitrag zur Effizienzsteigerung mobiler Arbeitsmaschinen. Entwicklung einer elektrohydraulischen Bedarfsstromsteuerung mit aufgeprägtem Volumenstrom. 2015
ISBN 978-3-7315-0339-2

Band 33 Rinaldo Arnold
Automatische Abstimmung der Sekundärseite eines dreiphasigen Systems zur berührungslosen induktiven Energieübertragung. 2015
ISBN 978-3-7315-0355-2

Band 34 Johannes Gültlinger
Kraftübertragung und Fahrbahnverschleiß durch Spikereifen. 2015
ISBN 978-3-7315-0358-3

Band 35 Thorsten Dreher
Energieeffizienz von Konstantdrucksystemen mit sekundärgeregelten Antrieben beim Einsatz in mobilen Arbeitsmaschinen. 2015
ISBN 978-3-7315-0377-4

Band 36 Steffen Kölling
Konzeptionelle Untersuchung zur Neigekompensation von Stromabnehmern. 2015
ISBN 978-3-7315-0387-3

Band 37 Michael Fritz
Entwicklungswerkzeuge für die Fahrzeugklimatisierung von Nutzfahrzeugen. 2015
ISBN 978-3-7315-0384-2

Karlsruher Schriftenreihe Fahrzeugsystemtechnik (ISSN 1869-6058)

Herausgeber: FAST Institut für Fahrzeugsystemtechnik

Band 38 Ralf Oberfell
Stochastische Simulation von Energieflüssen im Nutzfahrzeug. Ein einsatzorientiertes Bewertungs- und Optimierungsverfahren. 2015
ISBN 978-3-7315-0403-0

Band 39 Christoph Sturm
Bewertung der Energieeffizienz von Antriebssystemen mobiler Arbeitsmaschinen am Beispiel Bagger. 2015
ISBN 978-3-7315-0404-7

Band 40 Florian Netter
Komplexitätsadaption integrierter Gesamtfahrzeugsimulationen. 2016
ISBN 978-3-7315-0414-6

Band 41 Markus Springmann
Auslegung eines asynchronen Langstatorlinearmotors mit großem Luftspalt als Straßenbahnantrieb. 2015
ISBN 978-3-7315-0418-4

Band 42 Alexander Basler
Eine modulare Funktionsarchitektur zur Umsetzung einer gesamtheitlichen Betriebsstrategie für Elektrofahrzeuge. 2015
ISBN 978-3-7315-0421-4

Band 43 Hans-Georg Wahl
Optimale Regelung eines prädiktiven Energiemanagements von Hybridfahrzeugen. 2015
ISBN 978-3-7315-0422-1

Band 44 Jennifer Heck
Zur Simulation des Rad-Schiene-Verschleißes bei Straßenbahnen. 2016
ISBN 978-3-7315-0443-6

Die Bände sind unter www.ksp.kit.edu als PDF frei verfügbar oder als Druckausgabe bestellbar.

Karlsruher Schriftenreihe Fahrzeugsystemtechnik (ISSN 1869-6058)

Herausgeber: FAST Institut für Fahrzeugsystemtechnik

Band 45	Moritz Vaillant **Design Space Exploration zur multikriteriellen Optimierung elektrischer Sportwagenantriebsstränge: Variation von Topologie und Komponenteneigenschaften zur Steigerung von Fahrleistungen und Tank-to-Wheel Wirkungsgrad.** 2016 ISBN 978-3-7315-0452-8
Band 46	Philip Nagel **Entwicklung einer Betriebsstrategie zur Energierückgewinnung in hybriden Mehrverbrauchersystemen.** 2016 ISBN 978-3-7315-0479-5
Band 47	Matthias Pfriem **Analyse der Realnutzung von Elektrofahrzeugen in kommerziellen Flotten zur Definition einer bedarfsgerechten Fahrzeugauslegung.** 2016 ISBN 978-3-7315-0489-4
Band 48	Mohanad El-Haji **Ontologie-basierte Definition von Anforderungen an Validierungswerkzeuge in der Fahrzeugtechnik.** 2016 ISBN 978-3-7315-0496-2
Band 49	**9. Kolloquium Mobilhydraulik 22./23. September 2016 in Karlsruhe.** 2016 ISBN 978-3-7315-0573-0
Band 50	6. Fachtagung **Hybride und energieeffiziente Antriebe für mobile Arbeitsmaschinen 15. Februar 2017, Karlsruhe.** 2017 ISBN 978-3-7315-0601-0
Band 51	Fabian Schirmaier **Experimentelle Untersuchung und Simulation des Umformverhaltens nähgewirkter unidirektionaler Kohlenstofffasergelege.** 2017 ISBN 978-3-7315-0620-1

Die Bände sind unter www.ksp.kit.edu als PDF frei verfügbar oder als Druckausgabe bestellbar.

Karlsruher Schriftenreihe Fahrzeugsystemtechnik (ISSN 1869-6058)

Herausgeber: FAST Institut für Fahrzeugsystemtechnik

Band 52 Mathias Cabrera Cano
Neuronale Netze mit externen Laguerre-Filtern zur automatischen numerischen Vereinfachung von Getriebemodellen. 2017
ISBN 978-3-7315-0621-8

Band 53 Arwed Schmidt
Flottenbetrieb von elektrischen und autonomen Serviceagenten im städtischen Personennahverkehr. 2017
ISBN 978-3-7315-0633-1

Band 54 Katharina Knaisch
Untersuchung von Spulensystemen zur induktiven Energieübertragung von Elektrofahrzeugen. Vergleich von Topologien und Entwicklung einer Auslegungsmethodik. 2017
ISBN 978-3-7315-0630-0

Band 55 Frank Christof Stalter
Ansätze zur akustischen Optimierung von Reifen und Fahrbahnen für Elektrofahrzeuge unter Antriebsmoment. 2017
ISBN 978-3-7315-0645-4

Band 56 Steffen Rose
Modellbildung und Simulation von mobilen Arbeitsmaschinen. Untersuchungen zu systematischen Modellvereinfachungen in der Simulation von Antriebssystemen am Beispiel Bagger. 2017
ISBN 978-3-7315-0684-3

Band 57 Ulrico Peckelsen
Objective Tyre Development. Definition and Analysis of Tyre Characteristics and Quantification of their Conflicts. 2017
ISBN 978-3-7315-0713-0